Irrigation in Tropical Africa

Problems and Problem Solving

33913 (67) Irr.

Edited by W. M. Adams and A. T. Grove.

CAMBRIDGE AFRICAN MONOGRAPH 3

The Centre for African Studies was founded in July 1965 to facilitate inter-disciplinary research and teaching in modern African Studies in Cambridge. The African Studies Series is published by Cambridge University Press in association with the Centre.

This is the third volume in the Cambridge African Monographs series, · published by the Centre itself. The aim of this series is to publish occasional research reports, seminar and conference papers of importance to African Studies but which might not be available. This volume contains papers presented at a workshop and organised by the Centre in March 1983 on Irrigation in Africa, Problems and Problem Solving. Details of the monographs are given on the back cover.

A. T. Grove
Director
African Studies Centre

BRITISH LIBRARY CATALOGUING IN PUBLICATION DATA:

Irrigation in Tropical Africa.
   1. Irrigation—Africa   2. Irrigation—Tropics
   I. Adams, W. M.   II. Grove, A. T.

631.7'0967   S616.A44

ISBN 0-902993-10-0

Copies of this Volume, and details about other centre publications, can be obtained from the Secretary, African Studies Centre, Free School Lane, Cambridge CB2 3RQ, England.

Contents

List of Figures

List of Tables

CONTRIBUTING AUTHORS

Bill Adams,

Department of Geography,
University of Cambridge,
Downing Place,
Cambridge, CB2 3EN

John F.Aitken,

Soils and land resources consultant,
8 Rook Grove, Willingham,
Cambridge, CB4 5EZ.

Alan Bird,

Sir M.MacDonald and Partners Ltd.,
Demeter House, Station Road,
Cambridge, CB1 2RS.

R.Wayne Borden,

Minister Agriculture Limited,
Belmont, 13 Upper High Street,
Thame, Oxfordshire, OX9 3HL.

John Briggs,

Department of Geography,
University of Glasgow,
Glasgow, G12 8QQ.

Martin A.Burton,

Sir M.MacDonald and Partners,
Demeter House, Station Road,
Cambridge, CB1 2RS.

Michael K.V.Carr,

Silsoe College, Silsoe,
Bedford, MK45 4DT.

Richard C.Carter,

Silsoe College, Silsoe,
Bedford, MK45 4DT.

David Dent,

School of Environmental Studies,
University of East Anglia, Norwich,
NR4 7TJ.

Tom R.Franks,

University of Bradford,
Project Planning Centre for
Developing Countries,
Bradford,
W. Yorks, BD7 1DP

W.Jack Griffith,

Long Common, Debdin Road,
Newport, Essex.
(Consultant to Sir M.MacDonald and
Partners, Cambridge).

Dick Grove,

Director, Centre for African Studies,
Free School Lane, Cambridge,

Francine Hughes,   Department of Geography,
                          University of Cambridge,
                          Downing Place, Cambridge, CB2 3ES.

Cecile Jackson,   35, Orchard Street,
                          Canterbury,
                          Kent.

Melvin G.Kay,   Silsoe College, Silsoe,
                          Bedford, MK45 4DT.

Richard W.Palmer-Jones,   Institute of Agricultural Economics,
                          University of Oxford, Dartington
                          House, Little Clarendon Street,
                          Oxford, OX1 2HP.

David Potten.   Hunting Technical Services Limited,
                          Elstree Way, Borehamwood,
                          Hertfordshire, WD6 1SB

Julian M.Siann,   25 Howard Place,
                          Edinburgh, EH3 5JY.

Tom L.Wright,   ACT Consultants,
                          Sternfield, Saxmundham,
                          Suffolk.

ACKNOWLEDGEMENTS

We would like to thank Ann Clark and Clare Strowgler for their help
with organising the Workshop in March 1983 and Paula Munro for her
work on the text. Paul Howell kindly read all the manuscripts.

Bill Adams and Dick Grove.

FOREWORD

W.M. Adams and A.T. Grove

This book contains thirteen papers originally given at a workshop entitled Irrigation in Tropical Africa: Problems and Problem-solving, organised by the African Studies Centre in Cambridge. It was held in the University Centre in Cambridge on 24th and 25th March 1983, and its aim was to bring together field practitioners engaged in irrigation development in Africa, and academic researchers from various disciplines. Irrigation is expanding rapidly in Africa, yet there is still little published material on the new schemes, the expectations for them and their achievements. The workshop therefore had two purposes. First, to try to draw together a picture of the state of irrigation planning and development in different parts of Africa and to link together some of those engaged in this work. Second, to begin to open up discussion both among practitioners and between practitioners and academics concerning the factors which contribute to and limit the success of irrigation schemes in Africa.

African irrigation development is at an exciting and critical stage. Apart from a few celebrated early schemes, notably Gezira in the Sudan, and some relatively small-scale colonial and post-colonial irrigation schemes for example in Nigeria, Sudan and Kenya, formal irrigation development has come late to Africa. In the last decade, however, the pace and scale of development have increased, and projects have sprung up at all scales over much of semi-arid Africa. Many of these have involved the use of expatriate expertise and participation by, among others, British consultancy and contracting companies. Despite high hopes, and best efforts, these schemes have frequently encountered problems in implementation. Such problems have related particularly to the difficulty of integrating national objectives for agricultural development (and the established technical procedures for project survey, appraisal and management) with the needs, aspirations and attitudes of participating farmers.

Obviously the development of an irrigation project does not end with the solution of the technical questions of water delivery and engineering structures. Indeed, the real questions of the human aspects of scheme development and the complexities of scheme management are at that stage just beginning to appear. The Cambridge workshop was planned as a vehicle for the discussion of problems, and the solutions to those problems, at all stages of development.

We believed, and still believe, that these problems are solvable, and that in some circumstances irrigation can make a valuable contribution to agricultural and national development in Africa. But the problems are not trivial. There are serious difficulties with the introduction of irrigation, particularly perhaps with large-scale projects. It is only by meeting in this way, informally, to discuss irrigation with others experienced in practical, technical and theoretical fields, that solutions can be hoped for.

The workshop seemed to be reasonably successful in promoting discussion of this sort, realistically -based and constructively -orientated, although probably we heard too much about the problems and had too little time to consider solutions. Real solutions will not come quickly, but without frank debate they are unlikely ever to emerge. We hope that this book will now carry it to a wider audience.

The thirteen papers in the book are not divided into fixed groups, but they do address certain consistent issues. The first article, by Jack Griffith, sets the context by taking a broad look at some of the questions of equality, economics and politics which are being raised by irrigation development in Africa, on national and local scales. Griffith looks in particular at Nigeria. The second article, by David Potten, provides a profile of irrigation in just one country, Madagascar.

The next two papers look at soil survey, often the first stage of irrigation scheme planning. There are well-established set procedures for soil survey. Wayne Borden describes some of the practicalities that lie behind them, taking as an example a project in Kenya, while David Dent and John Aitken use evidence from the Bacita Scheme in Nigeria to question the usefulness of some of the established approaches.

Carter, Carr, Kay and Wright develop this planning theme by looking at the manpower needs of an expanded irrigation sector, again referring particularly to Nigeria, and Francine Hughes describes a second, often forgotten, element in irrigation planning, the provision of fuelwood on a scheme in Kenya. There follows a paper by Cecile Jackson which has become available since the meeting; it considers the contribution of women to a scheme in the predominately muslim Kano region in northern Nigeria.

There then follows a series of papers which look at specific problems of irrigation schemes, both in their development and running. Alan Bird examines the intractable nature of problems of land tenure associated with large schemes in northern Nigeria, Julian Siann discusses the labour requirements of new irrigation schemes as a factor in their rejection by farmers, and Richard Palmer-Jones analyses the deep-rooted causes of persistent low productivity on smallholder schemes. Two further papers look at farmers' responses to irrigation: John Briggs describes the logic of farmers' choice of crops in the Sudan and their rejection of suggested cropping patterns;Bill Adams argues that the introduction of irrigation on one northern Nigerian scheme brings uncertainties for farmers who respond in the way they have traditionally done to the hazard of drought.

It is apparent that the key to making schemes work is radically improved management structures and procedures. One element in this is training, and in the last paper Martin Burton and Tom Franks describe recent developments in irrigation management training using methods of simulation and modelling.

These papers cover a wide range of topics, but their coverage is not comprehensive. There are a number of notable gaps, for example indigenous irrigation and water use, economic analyses of established schemes, discussion of the relative merits of large and small-scale schemes, and an account of approaches to the environmental and disease aspects of irrigation development. Some of these were discussed in other papers or talks given at the workshop, but which unfortunately are not available for reproduction here. Godfrey Ackers and John Gowing spoke on engineering measures for the control of schistosomiasis, and in an evening session John Sutton presented some of the evidence for precolonial irrigation in the East African Rift. In addition, John Robson prepared a paper on the regional environmental basis of irrigation in Africa, Tina Wallace one on non-farming occupation on the Kano River Project in Nigeria, and Colin Taylor spoke about irrigation in northern Ghana.

This collection of papers is not comprehensive, but it does open up a number of fruitful avenues for further research and action, and it is based on contributors' practical experience and field research. We hope that it will make a useful contribution to more relevant study and more successful implementation of irrigation projects in tropical Africa.

IRRIGATION IN NIGERIA:"MADNESS OF FOWL, MARRIAGE WITH CAT?"

W.J. Griffith

INTRODUCTION

Large scale irrigation development started in Nigeria with the coming of the oil boom in the early 1970's. As the Banker's Trust put it in a monograph:-

During the 1970's Nigeria emerged as one of the most important OPEC suppliers of premium crudes. It gained a capacity to produce up to 2.3 million barrels per day which, when combined with sharply higher oil prices, completely transformed the economy. Oil revenues rose ...from 25% to 85% of Government receipts ... economic growth was rapid ... in 1980, Nigeria's foreign exchange reserves were still well in excess of the total debt ...

It was as though all doors had been opened to limitless development opportunity. Small wonder then that the many entrepreneurial spirits, awakened to the possibilities of acquiring prodigious wealth, began immediately to consider major development possibilities and especially to exercise their commercial acumen in asking themselves "What will be my share?"

IRRIGATION

Two major irrigation projects were conceived in the early 1970's in Nigeria, one - of 25,000 ha - at Bakolori in Sokoto State, and another - of 66,000 ha - north of Maiduguri in Borno State; work began on both in 1975. Though there was at the time sound reason for developing the water resources of the drier northern regions of the country (the Sahel droughts hit hardest in the years 1972 to 1973) there were also demanding political reasons for urgent action - the petroleum revenues had to be spread regionally, and the agricultural sector was the largest vote-provider. Furthermore the contracts for the extensive works needed must have offered substantial opportunities for continuing Federal Government patronage in the States. There can be little doubt that the political aspects of these developments must have quickly become as important, or more so, than the technical. This paper offers a very brief view of some of the major consequences of this on the ground.

Within this term "technical" the areas of greatest uncertainty have turned out to be the social and managerial aspects of the projects, and it seems quite clear now that the latent difficulties were never fully appreciated at the outset, at least in regard to Nigerian conditions.

It may be that the politicians' eyes had been over-much caught by projects in other countries which had been too quickly accepted as suitable prototypes for Nigeria. There was the Sudan Gezira project with its vast areas of irrigated land, its evident commercial and agricultural success; not so dissimilar a territory, one might at first sight think, to that in Borno State south-west of Lake Chad. Also there was the Tennessee Valley Authority in the USA, where the series of dams on the Tennessee River bore at least superficial resemblance to the series of irrigation possibilities on the Sokoto and Rima Rivers of Sokoto State. But in the Sudan there had been secure land tenure and administrative control - lacking in Nigeria - and a closer analysis of the circumstances in the Tennessee Valley would have shown up differences sufficient to invalidate any hasty comparison with the river valleys of Sokoto State.

FARMERS

One might have expected dialogue on irrigation development in Nigeria to have been opened with those who were already farming land suitable for new projects but in fact farmers seem largely to have been ignored - perhaps because, in northern Nigeria, authority is traditionally unlikely to take soundings of popular opinion before action. Since the historic administrative attitude has long been Islamic and hierarchial, one would not expect the Emir to invite consultations in the market place. During my touring of irrigation areas I can typically recall a District Head saying to me "The peasant farmers? They don't know anything!". And on another occasion, in trying to get some idea as to whether Fulani graziers had themselves any improvements in mind in development areas, I asked one of them what he most wanted in life. He replied "Beyond good grass, only the wind!" Even today, it seems, the further from the main road, the greater the acceptance of the provisions of the Almighty. The farmers, in fact, neither expected nor requested participation - but then they had no intuition as to what changes were afoot.

Nowadays, of course, there is much more discussion at lower levels in the Local Government Councils than hitherto and one might expect that here would be found more impetus towards development and improvement generally. In fact, in my experience, it is as yet seldom the case. The development urge is there, of course, but it is individual not corporate; and usually where issues of communal interest are raised the reaction is rather to withdraw such sensitive matter from public debate and to consider in private the questions of privilege, patronage and preferment that they are bound to raise. Under the circumstances it is not in the least surprising that farmers were so seldom consulted.

As consultants we have often tried to redress this defect, but one cannot go too far in doing so without appearing to encourage subversion, which could only be counter-productive - if not akin to treason. And where there exists really strong feeling about traditional issues there is anyway no chance of hiding the fact and, as will be shown, farmers do eventually have their say - if only by insurrection. In such cases, though, there typically exists only a brief moment for logical discussion before communication breaks down utterly, and there are left only two hostile parties facing each other across the barricades. After this, practical experience seems to indicate that only force remains. There exist no other simple early warning systems.

5

PUBLIC RELATIONS

Seeing how matters of this nature stand we have sometimes suggested to the Authorities that they take special steps to improve their public relations, and have explained in some detail the necessity of doing so. They have sometimes then moved toward acceptance in principle, but they may well have convincing enough reasons to act only slowly. Farmers, as I never tire of saying, are not fools; they understand resettlement procedures very quickly and, being entrepreneurs to a man, are only too quick to seize any advantage which development makes available. Where uncropped land is to be flooded, if the farmer hears that compensation will be paid for farms, new farms will spring up overnight. If buildings are to be counted for compensation in irrigation areas, one can find rural areas which quickly appear to become semi-urban. Where land is vacant, grazing rights can rapidly emerge. Accordingly all development Authorities will be wary and on their guard.

Where they do make a move towards informing the public, the approach can be sadly authoritarian. On one occasion a River Basin Development Authority (RBDA) instituted a "Public Enlightenment Committee". The title was significant, since public opinion was clearly instructed, not incorporated; this was, of course, an entirely traditional approach - much as the Emir would send his official messenger to call on the Village Heads concerned, and to tell them "the Chief says ... " On such occasions there could seldom be any give and take; one does not argue with one's spiritual mentor.

Nevertheless, it has become doubly apparent over the years that this lack of consultation with farmers has been one of the cardinal mistakes of some of the major irrigation projects. Whatever the detailed nature of the works proposed it can in the end be only the local farmers who, in whatever capacity, will farm the scheme. If they cannot fully understand the proposals and their implications, if their land is harshly expropriated, if they can see no early benefit for themselves, if they are not permitted to grow crops which they need for subsistence, if the work programmes do not suit their family requirements, if their trade outlets are unduly upset or even closed, then they will be disinterested, and the project is lost until matters are changed. One can only be astonished at the ingenuousness with which the Authorities have offended all these precepts, yet still expect the project farmers to be grateful to them.

LAND

One of the most troublesome aspects of irrigation schemes in northern Nigeria is that of rights in land. Part of the problem is due to the fact that until quite recently statutory rights were virtually restricted to the towns while most farmland was held under customary tenure. This worked well enough in colonial times but, with the increasing pressure of population, customary tenure is rapidly giving place to individual rights - effectively, of freehold. And now that large-scale agricultural developments are becoming commonplace, where Federal or State Governments require development land, the Law in effect requires them to acknowledge this, to acquire land formally, and to pay heavy compensation for the privilege. Executive Authority meanwhile tends to counter the trend by doing its level best to keep costs at a minimum.

The current Land Law in Nigeria is "The Land Use Act". Originally this saw life as a Military Government Decree in March 1978 - the purpose of the legislation being (in the words of Brigadier Yar 'Adua, then Chief of Staff, Supreme HQ) as reported by the present Oba of Benin as follows:-

(i) Traditional land-ownership had contributed a barrier to national development programmes and Government were having difficulty in acquiring land for its development projects;

(ii) Government could not be indifferent to a system where only the rich and powerful owned land;

(iii) Land racketeering and unending litigations in land transactions had become the order of the day.
                    (New Nigerian July 10 1982)

Under this law, which in September 1978 became entrenched in the Nigeria constitution, all Nigerian land was vested in the State Military Governor. As civilians took over, provision was made to transfer the powers from him to the "Military Administrator", but apparently not to the "Civil Governor" - consequently the judges disagree as to whether or not the law remains effective. The matter is now believed to be before the Supreme Court who will no doubt pronounce in due course.

Assuming the Law is still effective, it clearly has great relevance to any schemes for land improvement. For example, if a RBDA undertook to build a dam it would, under the law, be required to acquire the necessary land for project works, for approach roads etc., and for the area which would subsequently be inundated by the reservoir, and to pay compensation therefore. Before acquisition, the procedure set out in the law must be completed, the rights formally revoked by the (Military) Governor and the revocation "signified under the hand of a public officer duly authorized in that behalf ... and notice thereof shall be given to the holder". But I have to confess that, law or no law, never have I ever seen evidence that this has been done anywhere.

COMPENSATION

In such development projects compensation has normally been paid, though frequently after much delay; but it would seem that payment has often taken the form of a negotiated lump sum or lump sums, and it is often difficult to be certain exactly what are the components. The main problem here concerns compensation for land. The law says that where rights are revoked the holder of the land "shall be entitled to compensation for the value ... of their unexhausted improvements". Unexhausted improvements are defined as "anything of any quality attached to the land, directly resulting from the expenditure of capital or labour by an occupier ... and increasing the productive capacity ... "

In practice, the trend has sometimes been for Authority to say that "we are paying compensation for improvements only - not for land" and to attempt to pay a much lower figure than might be paid for land under statutory rights, and lower perhaps than its market value. At other times it has not been thought necessary to pay compensation at all for land improvements since, to take advantage of the irrigation system to be introduced, the plots have not been acquired, but simply reshuffled. The trouble is that when the natural irregularity of the existing farm layouts is exchanged for the constrained tidiness of a rectilinear irrigation field, the new plots are never the equivalent of the old - whether in situation, soils, convenience, size or even productivity. It needs some very rapid thinking to overcome this sort of difficulty with farmers, particularly when they were never consulted before the project took off in the first place.

CHANGED CIRCUMSTANCES

There is no doubt whatever that this uneasy situation troubles both farmer and even the Village Heads, and brings them with a jerk into a changed set of circumstances with uncertainties on every side. Hitherto land was always recognised as a common good held on behalf of the community by the Village Head. Should a newcomer arrive at the village in previous times it would have been correct for him to present himself to the Village Head and ask for a plot. If a plot were vacant the Village Head could give him one. Or should a man leave his plot and not return in reasonable time there would be no problem for the Village Head to reallocate it. No plot would be sold; if vacant, the plot reverted to the community.

Today, as recently put to me by a Village Head himself, "Land used not to be bought and sold; now, though, a Fulani who has settled for a while will - when he wants to move on - very likely call another farmer and say to him "buy this farm". If he is questioned, he will merely say he is farming there for his living. They none of them inform the Village Head! If they did, would I not stop them?!" Later in the discussion he made another significant comment: "Our wealth," he said, "is no longer in our cattle, but in our farms." In practice then, both farmer and grazier own farmland.

The traditional view is losing acceptance. The Village Heads, in varying degree, are losing their position as customary custodians of the village lands. Even the Emirs themselves are losing their position as arbiters in the disposition of lands and of the associated patronage. Instead there is a move toward something like a land registration which, though at present gradual and low key, is in future likely to become of more significance. Some of the more powerful farmers have decided to take out Customary Certificates of Occupancy from Local Governments; this is perfectly permissible under Section 6 of the Land Use Act, but if the practice persists it will inevitably reduce the general acceptance of customary land tenure, and will further remove the interests and authority of the Village Heads; also, Local Governments are ill equipped to act as land developers - a point considered further below. Others have bought land (particularly irrigated land on the new projects) and have drawn up agreements which

8

bear the signatures of all parties - including that of the Village Head. Frequently the sales include all economic trees, but common practice is then to rent out such plots less the trees. Commerciality is ever of increasing importance and land values are rising steadily.

In short, such has been the rate of commercial agricultural development that Land Law and Administration has not kept pace. The Federal Government is learning this the hard way - at Bakolori, for instance, the farmers' uncertainty led to a peasant revolt. Tired of petitioning for compensation for lands which had in effect been alienated for project works, in desperation they closed all project roads and would allow no contractors' vehicles to move until their grievances had been attended to. In response Government took a tough line, moved in mobile police, burned villages considered to be recalcitrant and shot a considerable number of dissidents - the Press said 14, if I remember right, but other independent journalists listed 126 and claim that the total dead exceeded three times that number.

LAND ACQUISITION

The lesson that Government seems to have drawn from Bakolori is that outright acquisition of project farmland for irrigation projects is normally too dangerous to be allowed to become standard practice. The Government stance seems now to be that land is not acquired but simply reshuffled so that each farmer may obtain water at his plot in the new irrigation network. For several reasons this position is anomalous however, since it is quite clear that legally the land given over in perpetuity to project works (dam, canals, roads etc.) must be acquired in any case. And since project works may take over up to 20% of the available project lands, the remaining area which will be given over to irrigation farmland will be only about 80% of that previously enjoyed by the farmers before the scheme began. In other words the new plots will be smaller. If it is accepted that the provision of water provides the makeweight, then this gives the farmer a lien on water supplies - for which the Authority would find it hard in equity to make a change.

More problematic is the thought that if Government's position is that it has not acquired the land but merely reshuffled the plots then it has forfeited any legal right it may have had to order farmers how to farm their plots, how to use their water, what rotations to use etc., etc. - in other words there can be no local control whatever, and farmers can only be enticed by commercial incentives, or perhaps by example, to undertake farming according to the project plan.

More problematic still, this means that the project management must, unless they can persuade (not order) the farmers to farm co-operatively, live for ever with small plots - with few possibilities of mechanization, low productivity and general inefficiency - not a bit according to the project design, and with little prospect of a commercial return. It means, of course, that the project capital (in the case of Bakolori, probably of the order of 500 million Naira) must be regarded purely as social. It may in the end help the farmer personally, but it is unlikely to be of great help to the nation's food supply. Naturally, too, it will be asked: "Was this money well spent?"

HIDDEN AIMS

One may argue that all this is a rather overly-legalistic view, and it is true that the law is a bit of an ass in these matters. But the real charge is bad administration coupled with a certain political dishonesty - project aims have never been fully clarified, farmers were not principals in development but pawns - and that the real aims were (and still are) hidden. I take it therefore that the real (and recondite) aim was never agricultural development first of all, but chiefly to maintain and extend throughout the country the patronage and position of the elite. If there were any pseudo-agricultural aims at all, it may have been the idea of providing the city dwellers with cheap food, but it does not seem likely that this aspect was ever formally examined.

Let me say at once that I don't see that these hidden aims are necessarily bad or even impractical; all Governments everywhere tend to do what they can to reinforce their own status and position, and to look after their supporters. But it cannot be denied that so far the agricultural outcome has been unnecessarily disappointing. In any case, however, all is by no means lost since through these large projects the Government has provided itself with substantial and good quality (if expensive) infrastructures that can still be made to do useful work provided the fullest attention is now given to their proper development and management.

THE RIVER BASIN DEVELOPMENT AUTHORITIES

However, this is a big proviso; land is not the only legal difficulty. The uncertainty of aims is reflected also in the laws establishing the River Basin Development Authorities (RBDAs). The first two of these RBDAs (Chad and Sokoto Rima) were set up in 1973 but the first substantive law governing the functions of the first 10 of the RBDAs was made in 1976. Beyond the functions of generally undertaking development of water resources in their area, with all that that implies, they were specifically enjoined to "develop irrigation schemes ... and to lease the irrigated land to farmers ... for a fee ... " However in 1979 a new law removed this provision and substituted the function "to provide water ... for irrigation purposes to farmers ... for a fee." Originally, it seems, therefore, that RBDAs were to be regarded as developers of acquired farm lands, but now as technicians, earning money from supply of irrigation water - a profoundly different status.

At first sight, even though it preceded the Bakolori riots, the change would seem to be legally right and proper in view of the post-Bakolori attitude of Government to land, but in practice the position of the RBDAs does not appear to have changed much. The Authorities are anyway the only bodies capable of establishing irrigation schemes and of leading the farmers towards making a success of them. Like it or not they have to manage; and they have in any case become (or they should have become) land owners.

It is commonly believed that the RBDAs were modelled on the example of the Tennessee Valley Authority; but the circumstances of the TVA

10

were very different and their outlook was much wider. I do not know of any RBDA in Nigeria that has its eyes fixed on "the proper use, conservation and development of the valley's natural resources" or on the need to "care for the economic and social well-being of the people" or "to provide agricultural and industrial development by bringing experts to the farm and people" and so on. But (Inshah Alla!) it could conceivably develop that way, given a political prod from the centre. I believe it would be best if RBDA objectives did develop in this way since it would concentrate their attention on the physical and social well being of their areas and gain them the goodwill of the people. As it is, I see the current approach as being counter productive - even dangerous.

I have always believed in the integrated approach - and that one must always remember that to every action there is a reaction; to every cause, an effect. It is important therefore that RBDA's should carefully examine the likely effect of their actions. I am thinking, for instance, of the several dams built for the retention of irrigation waters and the consequent total change in the river regime downstream; of available land being inadequate for enforced resettlement; of the flooding of large areas of prime fadama land upstream of the dam; so that the already successful farmers downstream may have irrigation in the next decade.

No, I am confident that much of the RBDAs' intention is good, and that the initial development potential is in principle correctly assessed but what is lacking is a proper appreciation of the time scales and the social effects - or, if you like, any attempt to take the people along too. All development should have sound popular support and where development involves social costs these must be tackled first. The experts must first go to the people, and benefits from development should first of all accrue to those adversely affected, not those at a distance. I believe this should be the acknowledged perspective for all Nigerian RBDAs.

NEW TIER OF ADMINISTRATION

There is a further problem; the establishment of RBDAs as autonomous bodies capable of holding land in their own name is to establish a new tier in the Governmental Administrative machine. Several agencies now have jurisdiction in a project area but the spheres of operation and responsibilities of each have not been clearly defined.

In my experience, RBDAs have for this reason been circumspect with those who farm land in projects they are developing. Local Governments, under the law, have the power to control and manage the land within the area of their jurisdiction; but in fact they appear merely to have the simple ability to facilitate land development only by making it available, and none to control the purpose for which it is used.

This sort of difficulty will inevitably lead sooner or later to confusion. I cannot see what can be done without first of all giving much more thought to the detailed responsibilities of all the bodies concerned. This will be no simple task.

If the Local Governments are quite unable in practice to manage
land development, the RBDAs themselves are seldom as well equipped as
they would wish. Partly this is due to the fact that they do not have
the recurrent finances available to enable them to employ the numbers
or calibre of staff that the job would require - although this lack
would be fully in accord with the Government position already outlined
that RBDAs are no longer to be land developers as they were under the
1976 RBDA Act.

THE WAY AHEAD

What can be done? Perhaps the only way ahead would be to bring in
from outside recognised experts in commercial agriculture. Often
enough, though, expatriate companies and their African clients have
different views of what they are buying and selling. The former sell
modernity and high technology on the assumption of an appropriate
culture to make use of it. The latter tend to buy goods down to a
price and ignore the trauma of the social transformation that
modernity inevitably requires. Then when the crisis comes, and the
forecast revenue fails to materialise, no-one is prepared to face
making the change of social gearing needed - and quite right too, for
forced change of that sort is seldom justified. Clearly outside
experts must carefully appraise themselves of answers to these
situations.

The only sensible approach for the Nigerian circumstances seems to
be frankly to admit in all but exceptional cases that for the time
being Government expenditure on irrigation is to be regarded as being
for social purposes only and that there must be at least a moratorium
on all expectation of economic return. Governments must then get
together with the farmers and recast the projects so as to give them
what they want and in the way that they want it. They must seek to
offer clear demonstration of the economic value of any updated
irrigation technique in the hope that the farmers will adopt it. If
the farmers don't want it, the probability is that they are right and
the demonstration wrong. If the farmer is successful then the RBDAs
must reinforce success. Gradually it may be possible to induce farmers
voluntarily to form cooperatives, which should enable the project to
be modernised further. In the end it may be that the final state of
the project will be something approaching that for which it was
originally designed - but I can imagine that all this could take
twenty years, perhaps longer.

PARTICIPATION

My greatest faith is in the character of the African farmers, for
whom I have the greatest respect. I am not surprised to find that
gaining their cooperation, especially on large schemes and where they
have been badly treated, has proved difficult because (in the words of
Gavin Williams) "peasants lack any clear evidence that a
transformation of their way of life (along capitalist or socialist
lines) will ensure their security, improve their well-being, and
extend their independence - and they find considerable evidence to the
contrary."

I shall leave it at that, and you will see that I consider farmer participation to be vital. However, I should add that the farmer, particularly the Hausa farmer, has already a shrewd insight into all that has been going on. He knows quite well that it really is "madness of fowl, marriage with cat" to move too fast, and that he cannot be pitchforked into unsuitable and unprofitable modernity. But he has also known for a long time that "the hand that has soup gets licked" and he will not give up hope of getting his just deserts from the Government planners.

IRRIGATION IN MADAGASCAR: A BRIEF REVIEW

David Potten

BACKGROUND

Madagascar is the world's fourth largest island (Figure 1). It has a population of 9 million in an area as big as France and Belgium combined (586,000 km). The economy is dominated by agriculture; cattle raising and rice cultivation being the two most important activities. Irrigated agriculture is of major importance. If the term is interpreted broadly, to include all cultivation involving water diversion and some water control, then over 850,000 ha, or 40% of the cultivated area (2.2 ha), is irrigated. If a narrower definition is used, limiting the area to that irrigated from permanent structures operated and maintained by government or para-statal agencies, then some 200,000 ha, 9% of the total cultivated area, are included.

Rice is by far the most important irrigated crop, occupying over 90 per cent of the (broadly defined) irrigated area. Rice is the staple food in Madagascar, with consumption estimated at 160 kg per person per year. Until 1972 Madagascar was a net exporter of rice. Since then, however, imports have risen steadily, because production has remained static while the population has grown. In 1982 some 350,000 t, around 15% of national requirements, were imported. These imports represented 16.5% of total imports, by value.

The other main irrigated crops are cotton and sugarcane. The former is cultivated mainly by smallholders. The latter is grown both by smallholders and by plantations, most of them run by para-statal agencies.

On the high plateaux of central Madagascar the cultivation of paddy rice dates back many centuries, to the migrations of settlers from Indonesia in the 10th, 11th and 12th centuries. Their legacy is still evident in racial characteristics, language, the practice of rice cultivation in tiered paddy fields painstakingly established along river valleys and plains, and even in some of the rice varieties that continue to be cultivated. The organisation of rice farming in the high plateau areas also shows the characteristics familiar in long established rice cultures. The traditional village level organisations (the FOKONOLONA) have a continuing role in the coordination of water distribution and the mobilisation of labour for system operation and maintenance.

Around the coasts of the island the situation is very different. Most of the irrigation systems on the fertile river deltas were established during French colonial times or during the first decade of Madagascar's independence (the 1960's). The schemes were financed by private investors or foreign aid. The early schemes concentrated mainly on the establishment of private plantations, whereas later projects involved introducing irrigation to dryland farmers or cattle-raisers, or inducing resettlement from the relatively overcrowded high plateaux.

14

FIGURE 1

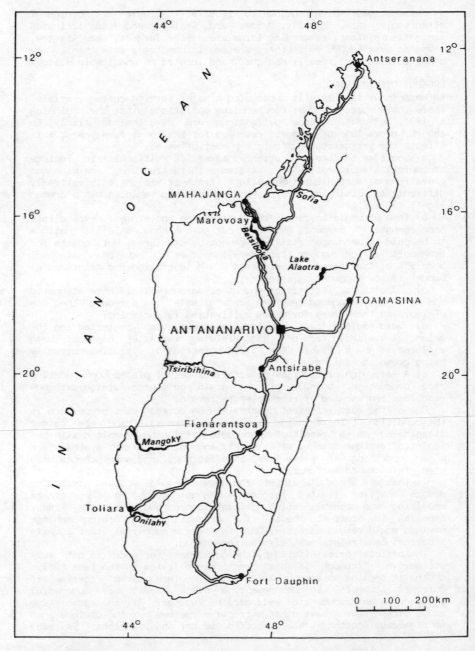

15

These scemes, which continue to receive the bulk of government attention, vary greatly. Some are well-planned, sophisticated irrigation systems, commanding large areas (the largest, Lac Alaotra, commands over 50,000 ha). Others are small (sometimes less than 100 ha under cultivation), poorly designed, and located on unsuitable soils.

PRESENT DIFFICULTIES

Madagascar is currently undergoing a major foreign currency crisis. There have been two debt rescheduling operations since the beginning of 1981, but debt service obligations are nevertheless likely to absorb over 40% of export revenues for the next five years. This affects the irrigation sector in a number of ways:

a) Shortage of foreign currency makes it difficult to replace mechanical equipment, for system infrastructure, maintenance operations or agricultural operations (tractors etc). It is extremely difficult to obtain spare parts, and immobilised vehicles are a common sight;

b) In a struggle to reduce the deficit on government expenditure 'non-essential' payments have been cut to the bone. Salaries continue to be paid to permanent staff who cannot be dismissed, but budgets for transport, office suplies, temporary labour, etc., have been slashed. Some 95% of the Irrigation Service's O & M (operation and maintenance) budget is now consumed by salaries;

c) Imports of fertilizers and other agro-chemicals have stagnated and no locally produced supplies are available. As a result the use of improved techniques for paddy cultivation is declining;

d) Seed multiplication activities have been neglected and the potential available from new high-yielding varieties has not been explored. As a result the yield potential of varieties currently being grown is limited;

e) A very high priority is nevertheless being placed on boosting rice production, because of the drain on scarce foreign exchange resources represented by recent import levels.

One of the most important constraints on current rice production is the condition of the existing irrigation schemes. Many are rather dilapidated as a result of inadequate maintenance combined with the impact of cyclone damage. Water supply and distribution systems are often operating at a low level of efficiency, and some schemes have completely ceased to function.

If this was the whole picture the answer would appear clear, to devote whatever limited foreign exchange resources there are to repairing and rehabilitating infrastructure and improving input supplies in order to boost rice production, as foreign exchange savings would result directly. Unfortunately a number of other equally important constraints must also be considered.

The official price offered by the government for paddy is not very attractive. Although it has been raised twice in the last twelve months it remains at around 50 to 75% of black market levels. If current attempts to suppress the black market are successful inducements to produce rice will decline further. At the same time many rice-growers have other attractive revenue earning options such as sugarcane, cotton, tobacco, coffee, pepper and tomatoes. In many areas the smallholder's strategy appears to be to grow enough rice for his family's needs, plus a small margin for seed and emergencies. His remaining efforts will be devoted to more remunerative crops, livestock etc. It is possible that the paddy price level that would be necessary to induce Madagascar's smallholders to produce enough to feed the growing urban population may be substantially higher than the import cost.

Madagascar is relatively sparsely populated, particularly around the coasts (4 to 10 persons per km), and rice development is frequently constrained by labour shortages. However in the present foreign exchange stiuation it is very difficult to promote mechanisation. Labour shortages are exacerbated in some areas by "FADY" days - two or three days a week are taboo, and no work is possible on them, even at peak transplanting and harvesting times.

Land tenure problems constrain development in many areas. On some schemes excessive fragmentation has led to the adoption of poor cultivation and irrigation patterns; on others there are wide disparities in land tenure: absentee landlords' land is either abandoned or cultivated by share-croppers. The effectiveness of agricultural extension is severely reduced by the same budgetary constraints that limit O & M activities. On the modern schemes smallholders were accustomed from the start to a system where all O & M activities were handled by the state. They are reluctant to start participating now in the management and operation of schemes.

Poor roads and a lack of vehicles (partly due to foreign exchange problems) make access to markets difficult; and, finally, law and order is tending to break down in some rural areas. In an atmosphere of insecurity smallholders are naturally reluctant to invest. In some areas farmers who used to cultivate with bullocks are returning to hand tillage as they are afraid to bring their animals near the rice fields - exposed areas where cattle rustlers may appear. Elsewhere labourers who used to migrate to areas short of manpower, either seasonally or for several years, now prefer to avoid unfamiliar territory, increasing the labour shortages referred to earlier.

PRESENT DEVELOPMENT EFFORTS

This gloomy description of present problems should not be interpreted as meaning that there is no further development potential or that no efforts are being made to improve the situation.

Official figures from the Madagascar Government's Directorate of Rural Infrastructure, which is responsible for most of the government managed irrigation schemes suggest that an additional 100,000 ha currently commanded by existing irrigation infrastructure are not being irrigated. Around the coast of Madagascar there are a number of river deltas with fertile soils which could be the subject of substantial further irrigation development - during the late 1970s the government itself was talking of bringing a further 100,000 ha under cultivation before 1990. Hence a doubling of the 'narrowly defined' irrigated area, to 400,000 hectares, seems well within the bounds of technical feasibility.

At present however all efforts are, very reasonably, concentrated on improving irrigation in currently or recently irrigated areas. A number of international aid agencies are involved in rehabilitation and repair operations, the main ones being the World Bank, UNDP/FAO, the European Development Fund, the French Fund for Aid and Cooperation (FAC) and the Caisse Centrale pour Cooperation Economique. The UNDP/FAO programme is concentrated mainly on the provision of minor concrete structures to some 170 essentially traditional small irrigation schemes (less than 200 hectares) on the high plateau, where many of the non-engineering problems are less severe. The programme is known as 'Operation Microhydraulique' and appears to be going well. The other agencies are involved in tackling problems on the larger irrigation schemes.

The main approach adopted can be characterised as 'full-blooded rehabilitation'. A typical such programme would include:

1) Civil and hydraulic engineering operations adequate to permit efficient operation of the system for at least 20 years and implementation of a rational water distribution operation (some redesign and reconstruction work may be involved).

2) Other infrastructure work necessary to permit effective exploitation of the scheme. Road rehabilitation is usually necessary, and some new staff housing, office or storage buildings are often required.

3) Substantial strengthening of the O & M organisation, in terms of equipment, recurrent finance, staff numbers and staff quality (training programmes in water distribution organisation and on farm water management are particularly necessary).

4) Simultaneous revitalisation of the extension services, again both in terms of recurrent finance and staff quality.

5) Improvements in the supply of production inputs - particularly fertilizers, credit and mechanisation services.

There is no doubt about the need for such comprehensive rehabilitation efforts. The costly civil and hydraulic engineering work will only be economically justified if the other 'side-issues' are tackled simultaneously. In the long term this is the only viable strategic approach.

In the short-term, however, the effectiveness of the strategy must be open to doubt. The limited resources of the Irrigation Service constrain its ability to effectively supervise the rehabilitation studies and work currently underway. Further, it is evident from the difficulties described above that an effective implementation of a full rehabilitation operation on a number of schemes would mean that very limited resources (particularly recurrent budget and foreign exchange resources) would have to be concentrated on these schemes, and that the rest of the sector may be even further starved.

Full rehabilitation can be justified for a few major "pilot" projects, on the grounds that the sooner a rehabilitation operation is launched, the sooner the necessary experience will be gained to permit wider implementation of rehabilitation programmes when national resource constraints are eased.

In the meantime what can be done for the 120 or so large schemes (500 to 10,000 ha irrigated) which will not be the subject of rehabilitation for at least the next five years, but are dilapidated and operating far below potential? A possible answer, currently being explored with European Development Fund finance, is to make brief reconnaissance surveys of each scheme, implement short-term repairs necessary to keep the schemes operational for at least five years (or bring them back into operation where complete failure has occurred), identify which schemes offer the best potential for rehabilitation in the medium and long term, and then commence any surveys necessary to permit rehabilitation when resource constraints permit.

CONCLUSIONS

It will be evident that the irrigation sector in Madagascar is in many ways atypical of African conditions:

a) A large traditional irrigation sector exists, with many centuries of experience and a 'rice-culture' more reminiscent of South-East Asia;

b) Even the 'modern' sector is relatively long established. Most of the large schemes were developed between 1930 and 1960;

c) The overwhelmingly dominant crop on both 'traditional' and 'modern' irrigation schemes is rice, mainly grown for home consumption;

d) The current development emphasis is not on the creation of new schemes but on repairs and rehabilitation of long established systems.

Nevertheless Madagascar is typical in one important way. The main problems that need to be overcome in irrigation development are not engineering issues but institutional, social, financial and economic constraints. The 'side-issues' are central.

SOIL SURVEY FOR IRRIGATION PROJECTS: A DISCUSSION OF TYPICAL PROBLEMS

R.W. Borden

INTRODUCTION

The methods and problems of carrying out natural resource surveys for irrigation in Africa are not unique. Survey methodology is standard whatever the end use and the problems generally stem from the same half dozen or so sources. The problems one can expect to encounter are also universal. For illustrative purposes I will focus on a project recently completed in Kenya.

This project is known as the Lower Tana Village Irrigation Programme (LTVIP) and is one of several such programmes being implemented by the Land Development Division (Small Scale Irrigation Unit) of the Ministry of Agriculture. The aim of LTVIP is to provide 50 to 200 ha irrigation schemes to villages along the lower reaches of the Tana River. This would allow 1 ha of irrigated land per village family with water being supplied by low lift pump from the Tana, Kenya's largest perennial river.

The project stretches along the river for about 100 km inland from the delta. Garsen, the ferry crossing, 100 km north of Malindi, is about mid point in the project area (Figure 2). The climate of the area is semi-arid to arid with annual average rainfall decreasing inland from the coast from about 1,000 mm to 500 mm. Rainfall is highly variable. There are two pronounced dry seasons: December to March and June through October. Temperatures are high throughout the year and average annual evaporation exceeds 2,000 mm.

The surveyed area covered the floodplains and levees of the Tana River as well as a narrow fringe of the adjacent terrace. The survey area is generally flat (slopes less than 2%) with the river levee lands being slightly above the normal flooding level of the river, and the basins just below it. A number of abandoned river channels and levee ridges occur in the flood plain. There are few meso but significant micro relief features, the former being slight changes in elevation (generally less than 2%), the latter being mainly gilgai features formed by swelling and shrinking of the clays, particularly in the basins. Meso relief features are the levees associated with the present and past river channels, and remnant sand ridges and fragments of terrace. One or two erosional remnants lie in the project area. Terraces rise abruptly away from the flood plain and generally lie ten to fifteen metres higher. A number of coarse textured (sandy) ridges occur within the terrace area. These usually appear as parallel ridges of up to 20 m high, elongated and oriented in a NE-SW direction. These ridges are most common south of Garsen.

In geological reports the area is mapped as recent alluvium with bands of older sand and clay ridges. These ridges were deposited during various oceanic phases and become progressively younger towards the coast. Recent deposits in the area consist of sands, muds and silt deposited during the biannual flooding of the Tana River. The origin of much of this material is in the highlands to the west and

FIGURE 2   PROGRAMME AREA LOCATION MAP

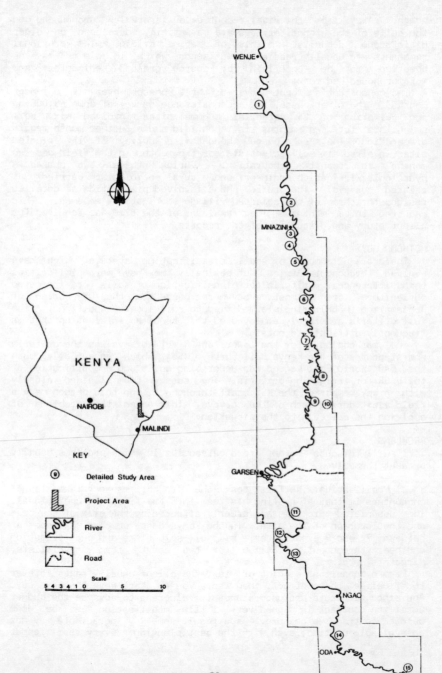

north. Away from the river recent deposits are less pronounced and the soils of the higher terraces are essentially developed on older Pleistocene deposits, mainly of marine origin. The erosional remanants are usually reddish brown sands which in places may be observed to be underlain by coarse grained sedimentary and metamorphosed bed rock.

The consultant's input covered three phases over 15 calander months. The first phase, the reconnaissance survey of some 65,000 ha and selection of 25 potential scheme sites involved six soil scientists for two months in the field and a further month report preparation. The second phase, the detailed survey of 15 selected sites involved six soil scientists for two months in the field and two months in report preparation. During these two phases a hydrologist/irrigation engineer, and a rural sociologist carried out related surveys. The third phase involved preparation of detailed designs for three of the selected village irrigation schemes. This required three months. The remainder of the time was required for client study and discussion of the reports.

METHODOLOGY

There are a number of soil classification systems which have evolved throughout Africa, but basically they have more similarities than differences. Criteria for classification of soils for irrigated agriculture are almost everywhere based on the US Bureau of Reclamation (USBR) principle modified to suit local conditions. The full criteria are rarely employed as this requires information that is frequently difficult to obtain.

The methodology for the Lower Tana River surveys was the criteria and standards of the Kenya Soil Survey (KSS) which essentially uses the FAO World Soil Legend for describing soils and a modification of the USBR for irrigation criteria. The survey team liaised closely with a key member of the KSS staff throughout, and this person made a field visit during each of the surveys. This eased the problems of applying the criteria to the situation.

PROBLEMS

I will discuss seven broad categories in which problems usually present themselves.

1) Mobilisation: The first task the project manager/team leader encounters is that of getting his team into the field and operational. The associated problems are greatly affected by the size of the team and the location of the area. In the Kenya case we had a team of between 7 and 9 expatriates – six soil scientists and one irrigation engineer throughout, plus one or two short term specialists, directors, etc.

From a consultant's point of view, the client usually takes forever to finalise the contract, then wants you to start work the next day. And since directors are understandably reluctant to advance the budget until the contract is signed very little mobilisation can be done before contract signature. As a result some key ingredients may not be available on time, even with the best planning. Every team leader can tell you a story or two of problems created by hasty or faulty planning and pre-project preparation, or of mobilising too quickly. Fortunately for the Kenya project this aspect had been handled well. There was adequate budget for equipment and the suppliers were able to provide it within the mobilising period. For accommodation and messing facilities we contracted a former big game hunter to set up and run the field camp. He also looked after general maintenance of our vehicles. This is by far the easiest way to solve the most important consideration: food and shelter. The client provided the team with an office block. We were also able to hire field labourers in Garsen, which meant that additional camp facilities were not required.

2) Aerial Photography/Remote Sensing: This is frequently a problem area. It was for us. One aspect that few people seem to be aware of is that there are very few days in the year that are ideal for taking aerial photographs. These are days when, literally, there is not a cloud in the sky. In Kenya, January and February are the best months. Our project was to begin in January but various delays meant it did not mobilise until July. Even though this is in the second dry season, there are very few flying days in the project area at this time of the year. As a result we found ourselves in the field without our photography; the plane standing by to fly at the first opportunity. What was to be done? The only photography available was of extremely poor quality and very much out of date. We found it impossible to locate roads, villages, etc., as these had changed location following particularly severe flooding in the mid 1970's. Our KSS liaison informed us that 1:60,000 photography had been flown the year before but on checking we found that it was still not available. The Survey Department did have one set of prints and agreed to photograph and reproduce the prints we required. The results were not ideal but better than the older photography. We were now able to find our way from place to place and start marking the locations of our observations with a degree of accuracy. Eventually we did manage to fly the new photography (scale 1:45,000) and had the prints for the last week of field work.

For the second phase, carried out in the January/March dry season we had a second set of photography flown, this time at a scale of 1:7,500. This was flown the last few days of December under ideal flying conditions. At this time the grass, mostly Hyparrhenia spp, was up to 2m high over most of the area and slightly taller in the lower areas. This presented some problems in aerial photo-interpretation, especially since much of this grass had been burnt by the time we arrived in the field. Further problems were encountered in preparing orthophoto maps from these photos since the electronic sensor was working on a surface somewhat above true ground level and one that tended to mask variations in topography. The task of co-ordinating this mapping was further affected by the fact that spot heights for the orthophoto mapping had to be done after the photography was flown which meant that the surveyors frequently could not measure the actual grass height as it was no longer there. In retrospect, the air photography for the orthophotos should have been flown in February/March in this area when most of the tall grass has been removed from the land surface. This would also have given time for the surveyors to have located and marked spot heights on the ground. However, such a flying schedule would have again put the soil surveyors in the field without adequate photography.

3) Equipment: The purchase and delivery of equipment should be a fairly straightforward activity but frequently problems are encountered in the chain, particularly at the receiving end. The equipment we ordered was assembled and dispatched with timely efficiency; the problems arose in Nairobi. Not that Nairobi is unique, I should think it is typical. Two days before we were to go to camp I discovered that the equipment was at the airport, it had been for 10 days. After much paper work, signatures, etc., I got the necessary clearance on the third day only to find that one of the three parcels was missing, or rather half of it was. The shipper had bound two boxes together and these had separated, and only one had the label on. I managed to convince the customs people that part of the package was missing and to get permission to look through the warehouse, a very large building. Fortunately after about half an hour I spotted a package that looked right; it was.

4) Local Politics: The soil surveyor does not usually get very involved in this though it can happen. In this particular study we found it difficult to flag our observations and the orthophoto survey crew working in the area frequently found their bench marks dug up. The reason behind this involved the local dispute over land use. Traditionally two groups of people use the river and the flood plain. One group is composed of fishermen and farmers who are settled in villages along the river bank. The other group is made up of semi-nomadic herdsmen who graze their cattle over the wider area during the rainy season and move into the flood plain during the dry season. The herdsmen saw the irrigation schemes as cutting them off from the river and dry season grazing and while they watched at a distance by day, they moved in at night to remove markers. This did not inconvenience us unduly but clearly pointed out the need to include all land users in development plans. In the LTVIP only a small percentage of the land will be developed for irrigation and should not affect adversely the grazing requirements of the herders. In fact drainage water from the schemes may even improve dry season grazing.

5) Terms of Reference: The TOR for the Kenya project were well written compared with many I see. Some appear to be quotes from text books, others, very vague with no thought for the detail required to achieve the end result. Unless considerable thought and planning has been put into setting out the terms of reference for irrigation development, a number of problems can arise. From the land classification point of view, considerable expense can be entailed by unclear rationale or requirements.
For example, the TOR for the Kenya project required a detailed reconnaissance survey of 65,000 ha to identify about 2,000 ha (in 50

to 200 ha blocks) of irrigable land near 25 villages. This was then
followed by a detailed survey of 1,800 ha in 15 parcels. Land
suitability maps were to be prepared for both surveys. Since the
stated aim of the project was to identify suitable irrigation schemes
for 25 villages, all located in a relatively uniform river plain, this
could have been achieved by conducting a very broad reconnaissance
survey of the larger area and then launching straight into the
detailed survey of the selected sites.

The average observation density required for both the
reconnaissance and the detailed survey was far too dense for the
stated objective and the land resource. The reason for the detail
specified was that there was a second objective. The detailed
reconnaissance was also to be part of the national survey grid. That
this survey was to serve two, maybe three, requirements presented no
specific problems per se for the surveyor. The cost, however, was
added to the specific irrigation project rather than apportioned to
the other users. This may not be a major consideration when aid money
is paying but it does distort the economic considerations and would
put an unjust burden on the villagers if they had to bear the full
costs.

6) Irrigation Suitability Criteria: In theory, any soil can be
irrigated, especially when any number of man-made modifications are
employed and if expense is not a limitation. Modifications include
addition of fertilisers, installation of subsoil drainage, development
farming systems which increase or decrease infiltration, and land
levelling to make uniform slopes or terraces. In practice, very
little amelioration can be done at the intermediate and low technology
level because the farmers normally cannot bear the input costs. Thus
the criteria for assessing land suitability for irrigation has to take
a "best to least best" approach in the present state and "if such and
such is done" approach to the potential for future state.

Land levelling is one of the easier modifications but will not
always improve the irrigability of the land. Where soils are shallow
or have saline or sodic subsoils levelling will expose subsoil
material and even bedrock in some parts of the field while increasing
the depth of topsoil in other parts,resulting in non-uniform growing
conditions, at least in the short term. Internal drainage is required
to keep a balance of salinity in the profile. Conditions have to be
such that there is a net downward movement, and ideally a flushing of
salt out of the profile. Fertilisers and other farm inputs can be of
little benefit if other factors are restricting.

While soil profile characteristics are important there are many
other factors that may have even greater influence on irrigation
development. Many of these factors are not in the realm of the oil
scientist. Political considerations and proximity to adequate water
supplies almost always dictate the location of the scheme, so that by
the time the soil surveyor arrives on the scene the site has been
selected and boundaries drawn. This can mean serious problems if the
land does not meet the criteria for irrigation suitability.

On the Lower Tana neither political nor water proximity criteria
were restricting to the soils team. It was the soils themselves. The

river flows through an old marine plain whose soils are predominantly calcareous clays and sandy clays with strongly developed coarse columnar and prismatic structure and vetic characteristics. They are typically highly saline and sodic, with poor drainage and very low permeability and infiltration rates. Over the years the river has cut into the marine plane and built up its own fluvial deposits. These too are predominantly clays with strongly developed prismatic or block structure with very low permeability and infiltration rates. Some of these profiles are also saline.

Land classification criteria for rice and for upland (dry foot) crops was set up covering topography, vegetation and several soil profile characteristics. Of the nine critical factors examined the principal factors of determination were soil texture, electrical conductivity and drainage.

7) Client Consultant Relationships: One way to ease or exaggerate problems is the way in which the client/consultant relationship is managed. A good relationship between the two allows for close liaison and awareness on both sides. A totally distant, non-contract relationship is a recipe for much frustration

On the Kenya job, we liaised very closely with representatives from the KSS and the client. These representatives did not work with us but we met with them frequently and they made a field visit. They knew our work situation and thus when problems arose we could discuss solutions and get approval for remedial action. Thus when the report was submitted, very little discussion or alteration was required.

SUMMARY

In the foregoing I have used the Lower Tana Village Irrigation Programme to illustrate some of the typical problems encountered when carrying out soil surveys for irrigation projects. There are many more areas in which problems could, and often do, occur. Space has not allowed me to touch on the importance of selecting socially as well as professionally compatible personnel, especially when the project is carried out in a remote location over an extended period of time, nor on the problems attached to obtaining skilled and semi-skilled local manpower, of obtaining vehicles and other essentials in countries suffering acute foreign currency shortages and of maintaining equipment under severe working conditions in remote locations. Local cultural practices and work ethics can also be sources of problems for expatriate survey teams, especially on projects that require short and quick inputs that do not allow time to understand and settle into the local scene. Any reader could add several paragraphs from his/her own experience. Each project has its own unique problems to be solved. These all add together to conclude that developing the academic theory is only a small step towards completing the job. Equally important is a pragmatic ability to adapt and devise in order to get the maximum information within a set budget and time span.

SOIL SURVEY FOR IRRIGATION: A CASE STUDY OF BACITA ESTATE, NIGERIA.

David Dent and John Aitken

INTRODUCTION

Has soil survey become a ritual? This question is addressed to both fellow soil surveyors and to the users of our surveys. What useful predictions can soil surveyors make from all the data they conventionally gather in the course of a soil survey? How much of this data do users need and what information do they really want from us? This last question is the most important one and it has to be clearly answered if the time and effort and funds expended in soil survey are to be worthwhile.

The soil surveyor, in cooperation with a team of other specialists, is required to do four things:

1. Locate and map land that will be responsive to irrigation,
2. Evaluate land suitability for irrigation, either qualitatively or in terms of benefits and costs,
3. Identify physical hazards or problems of soil and water management.
4. Provide guidelines for management that will avoid hazards and ensure efficient use of water and other inputs.

If all this information is to be used effectively it has to be directed to the people who can use it and it must be intelligible.

Irrigation involves supplying water in unnatural volume at unnatural times. It changes the relationships between land, water and production. The objectives of soil survey are to predict what will happen to the water and how the crops and environment will respond to the kind of water management that will be applied.

In tropical Africa, irrigation water is a scarce resource that should be used efficiently. Large-scale irrigation requires a sophisticated system of management: maintenance, water budgeting and other agricultural inputs, credit, storage, processing, transport and marketing. All of these are needed at the right time. From such a big investment in people and money an appropriate political and economic return is expected.

LAND CHARACTERISTICS

The physical characteristics of the land that mainly affect irrigated crop production are climate, the water requirements of specific crops, potential water supply and water quality, relief and soils. In any given project area, climate, water requirements and water supply are effectively uniform. The land characteristics that vary spatially over the area are relief and soil-water relationships.

Relief influences surface hydrology including flood and erosion hazards, surface drainage requirements, methods of irrigation, and land levelling and shaping requirements. Land survey is technically easy. We can see what we are surveying both in the field and on air photos, and topographical interpretation is straightforward.

Assessment of soil-water relationships requires the acquisition of data on the rate of infiltration of irrigation water, available water capacity, leaching requirements, the rate of drainage through the soil and the fate of the drainage water in terms of the local water table and the landscape as a whole. The last is crucial if salinity problems are to be avoided. All these characteristics are related to the texture, thickness and sub-surface topography of the sediments (especially any layer through which water percolates either very slowly or very rapidly) and to any interface there may be between contrasting layers in the soil that will impede rooting, drainage or cultivation. The soil surveyor is responsible for mapping at least the upper 1.5 m and possibly for greater depths if there is no drainage specialist in the team. Survey of these characteristics is technically difficult, especially in alluvial landscapes where sub-surface layers are complex and not obviously related to surface features. There is no alternative but to dig for the data.

Accurate measurement of soil characteristics takes a lot of time. Measurement of infiltration capacity or saturated hydraulic conductivity at a single site may take all day; a basin leaching test will take several days; realistic determination of available water capacity takes several weeks in a soil physics laboratory. For many soil characteristics, tedious field measurements are essential because surrogate estimates or laboratory tests are unreliable. For example, there can be a 2- to 700-fold difference between the field measurement of saturated hydraulic conductivity and a laboratory determination on an undisturbed sample.

SURVEY DESIGN

The surveyor has a choice of two strategies, either to survey soil characteristics on a grid or to map the soil using geomorphology as a guide.

The soil characteristics normally recorded in the course of an intensive grid survey can be measured rapidly in the field. They include the elevation and slope at the site, soil texture profile, significant pedological features such as calcrete or plinthite layers, salinity and pH. These data are used to interpolate between point measurements at a much lower intensity of infiltration, permeability, leaching requirement and water retention.

Soil mapping on a basis of geomorphology involves mapping distinctive landforms such as alluvial fans and river terraces. (In the first instance, mapping is based on stereoscopic interpretation of air photos). The surveyor then builds a conceptual model of the structure of each landform, based on selective field measurements (for example along transects) and his understanding of the sedimentological and hydrological relationships within the landscape as a whole. If the surveyor is a good geomorphologist and hydrologist, this strategy can provide much better value for money at the project location and feasibility stage than a grid survey. It can also serve as a basis for a do-it-yourself survey kit of critical soil characteristics, mapping units and appropriate sampling intensity for more detailed survey by the farmers or local technical staff (Northcote, 1984).

Whichever of the two strategies is adopted the surveyor usually produces a soil map. His mapping units, whether individual soil series or soil landscape units, are the carriers of the surveyed information. Data from measurements of, for example, available water capacity for one or a few sites in a mapping unit are assumed to apply to all sites within the same mapping unit, although there is a range of variation of that property within the mapping unit which should be specified. Alternatively, where the sampling density is adequate, the value of a land characteristic at a point can be estimated by using a weighted average of values measured at nearby sampling points, a process called 'kriging' (Burgess and Webster, 1980).

INTERPRETING SURVEYS

Two complementary kinds of interpretation of soil surveys are usually needed: the first involves transfer of technical experience from other areas with similar soils; the second involves economic evaluation of the different mapping units.

Experience of the behaviour of similar soils elsewhere allows hazards associated with the proposed system of land use, such as salinity, to be identified. Design standards can also be specified to avoid the hazards or overcome other limitations on production by indicating the appropriate method of irrigation, the volume and frequency of water application, drainage and land shaping requirements. Guidelines for the management of each combination of cropping system and soil mapping unit can be provided by matching the physical requirements of the crops with the limitations imposed by the land and secondly, recommending ways of overcoming these limitations by engineering measures and appropriate water management. This matching is carried out intuitively in collaboration with the agronomist and engineer.

Land evaluation involves assessing the suitability of the different mapping units for irrigation and their relative suitabilities for a range of alternative crops and systems of management. Such economic evaluation involves a comparison between benefits and costs. Money values must therefore be assigned to the capital costs of the irrigation works, maintenance costs, the variable costs of seed, fertilizer, fuel, water and labour, and finally social and environmental costs and benefits.

Clearly a great deal depends on the soil surveyor's prediction of crop and land performance. In fact, he usually has little data on the relationships between performance and the physical characteristics he has surveyed.

Evaluation usually assumes that the number, severity and ease (cost) of correcting limitations determine the potential of each mapping unit. There are two approaches: land is more suitable if either the same yield can be produced with less cost, or it yields better with the same cost. Somehow criteria are adopted to rate limitations and weigh them against each other. We do not know any reasonably consistent way of comparing or adding together the effects of different physical characteristics such as available water capacity and adequacy of drainage. Usually it is done intuitively.

Production is estimated from an optimum yield figure, assuming no

29

limitations and good mangement, and then this figure is downgraded for each mapping unit according to the severity of its limitations. There are four inter-related defects in this procedure. It is wrong to assume good management. For various reasons such as budgetary constraints, lack of spare parts and labour problems, management capability in tropical Africa is usually limited and its efficiency is nearly always overestimated when irrigation schemes are formulated and appraised. Secondly, inadequate weighting is given to the type of limitation that exists relative to the ability of management to deal with it. Thirdly, because a soil surveyor rarely returns to the area he has mapped to see for himself how his mapping units have performed and how management has coped with their limitations, country experience is not accumulated in a way that would allow it to be applied in the evaluation of new areas. Finally, comparison of the ease and costs of coping between different kinds of limitations, say a serious drainage problem on a clay soil and supplying additional water to a sandy soil, is very difficult to make at the feasibility stage, especially if the two mapping units concerned are far apart.

WHAT SURVEYS CAN AND CANNOT DO
   The users of soil surveys want to know:

   1) Which land within the proposed area is suitable for the irrigated land use envisaged, bearing in mind water availability and the technological capability of the farmers?
   2) What inputs are needed for each different kind of soil so that it can achieve its potential?
   3) What will these inputs cost?
   4) How far will the repayment capacity of the project support the development and continuation costs?

   Other questions such as farm layout, water distribution and detailed development are of secondary importance once the overall feasibility is established.
   The soil surveyor can certainly say which lands are suitable or not suitable for irrigation. He can also, from his soil survey data, give management guidelines for each mapping unit. What he cannot do is to foretell the effectiveness with which his recommendations will be adopted, unless he has experience of similar schemes in the country, which is unlikely. Thus he cannot accurately predict how mapping units will yield, either absolutely or relatively. It follows then, that to spend time trying to apply suitability ratings of numerous categories (e.g. FAO 1976) is illogical and likely to be misleading. It is also unnecessary because users are interested in the profitability or production from the whole area. They can accept a range of yields so long as they know a representative and sustainable average, and they are unlikely to be prepared to develop the project area piecemeal to pick out the assumed best and leave the worst parcels.

BACITA, A CASE STUDY.
   This view of the interpretation of soil survey has been developed

by Aitken (1983) following an analysis of yield data for sugar cane from Bacita Estate, Kwara State, Nigeria (latitude 9 N, longitude 5 E).

The yields of irrigated plant cane from 5,000 ha over ten years were analysed with respect to fourteen soil mapping units, cane variety, rainfall/evaporation pattern and method of irrigation. Each factor was considered both alone and in combination giving 50 permutations. For each permutation the average yield of each mapping unit has been calculated and the units grouped into sets if there is no significant difference (p = 5.0%) between the yields. Figure 3 shows the location of the mapping units relative to the geomorphology and stratigraphy, and Table 1 describes the soil mapping units.

Figure 4 shows the grouping of the mapping units into sets for each of the 50 permutations, and Figure 5 the same information in simplified form. It shows that in 46 out of the 50 permutations there occur no more than three sets. Furthermore, the mean and range of yields in these sets are sufficiently similar to allow definition of three groups A,B and C, such that in 44 such permutations each group contains no more than one set. The boundaries between the groups, placed to minimise overlap of sets between groups, are at 84 and 73 tonnes of cane per hectare.

Table 2 classifies the mapping units according to the number of times each falls into group A, B or C:

    Class 1   normally falling into Group A;
    Class 3   seldom falling into Group A and often into Group C;
    Class 2   intermediate between classes 1 and 3.

These categories denote respectively the best, worst and average soil mapping units at Bacita, based on their performance under irrigated cane over ten years.

There is a broad correlation of yield with geomorphological units. The higher-yielding soil mapping units include those of the Niger floodplain and the Oshin alluvium, while the poorer-yielding mapping units are those on heterogeneous fan deposits.

Table 3 shows the various evaluations of the soil mapping units at Bacita made by soil surveyors and agronomists before and during the estate development. Comparison with Table 2 shows that a number of wrong assessments were made. For example, unit E (Egbungi), on which the Estate was originally founded, has performed relatively poorly, but N (Nebung) has exceeded all expectations.

Aitken has established that potentially (assuming adequate water and that drainage and other limitations are removed) the yields of E and N are similar. However, their actual yields reflect the effectiveness of management in overcoming the limitations, and are significantly different. Thus Egbungi, a clay soil with a drainage problem, yielded within 75% of its potential only 4 years in 10. Nebung, a sand with low inherent fertility but adequate water and reasonable drainage, yielded 75% of its potential in 9 out of 10 years.

At Bacita, with surface irrigation, soil mapping units fall into one of three categories and yield on average between 62 and 82% of

FIGURE 3 Geomorphology and stratigraphy, with related soil mapping units, Bacita Estate.

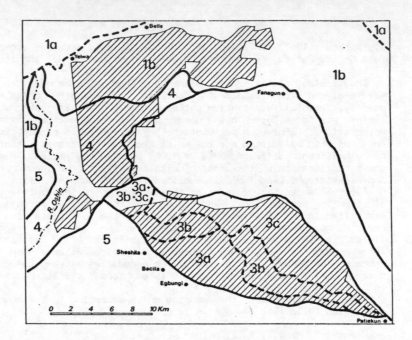

*for their distribution see soil map

GEOMORPHOLOGY		MAPPING UNITS OF ESTATE *	STRATIGRAPHY
1 Niger floodplain	1a - channels and sandbanks		
	1b basins and levees	1	Deep clays
		2	Deep clays, and shallow clays over sand
		3	Deep clays and clay loams
		4	Deep and shallow clays, loams and sands
2 Fanagun sandbank			
3 Alluvial and colluvial fans	3a coarse	R	Loam over sand
		Sh	Deep sand, and sand with clay lenses
		N	Sand over loamy sand (only in 3a + 3b + 3c)
	3b fine	E	Deep clays
		Ep	Deep humic clays
	3c variable	Eb	Clay over sand, and stratified clay and sand
		Br	Sand over clay
4 Oshin floodplain		B	Clay, sand below 1.0m
		Bs	Stratified clay and sand, over sand
		O	Deep clays
5 Nupe sandstone and old terraces			

	Soil mapping units		Description						
	Symbol	Soil series/complex	Morphology	a	b	c	d	e	f
Niger alluvium	1	complex of N 21 and N 22 series	Deep black and dark grey mottled clay and silty clay with moderate structure; poorly drained.	72 53	4.1 4.2	8	35 23	60 76	ts ss
	2	complex of N 21, N 22, and N 23 series	As above, but including soils overlying sand below 60 cm; poor and imperfectly drained.	71 68 1	4.0 4.0 5.3	8	36 37 3	57 67 70	ts ss (clay) ss (sand)
	3	complex of N 21, N 22, and N 12 series	As 1 above, but including light grey mottled clay loam and silty clay loam soils with moderate structure; imperfectly and poorly drained.	as 1 above, & also 44 35	 3.9 3.9	 6	 21 17	 53 62	 ts (N 12) ss (N 12)
	4	complex of N 12, N 13, N 21, N 22, and N 23 series	Variable complex of poorly drained deep and shallow clay and moderately drained sandy and loamy soils.	combination of 1, 2, and 3 above					
Oshin alluvium	B	Belle series	Thin loamy topsoil over light brownish grey mottled clay, with small sand pockets, overlying sand below 1m; imperfectly drained; moderate structure.	36 25 6	4.3 4.8 4.3	5	25 5	69 78	ts ss ss (sand)
	Bs	Belle series, sandy	As Belle above, but with larger sand lenses or sandy profiles; moderately well drained.	25 4 1	4.7 5.4 4.9	5 2	20 13	72 94	ts ts (sand) ss (sand)
	O	Oshin series	Thin loamy topsoil over massive grey mottled clay; poorly drained.	66	4.1	7	32	68	ts
Fan deposits — coarse	R	Rafia series	At least 30 cm of grey sandy loam to clay loam over grey structureless coarse sand; high water-table; poorly drained.						
	Sh	Shigo complex	Stratified clay and sand, and also deep grey mottled coarse sand; high water-table; poorly drained.						
	N	Nebung series	30 cm of grey sand over deep grey loamy sand; high water-table; imperfectly drained.						
Fan deposits — fine	E	Egbungi series	Up to 25 cm of grey mottled silt loam over moderately structured dark grey mottled clay and sandy clay, with small sand pockets in subsoil; poorly drained.	56 44 64	5.4* 6.0* 3.9	 2	20 20	 62	ts ss ts (T4)
	Ep	Egbungi series, peaty	As Egbungi above, but with a black humic profile and organic topsoil; poorly drained.						
Fan deposits — mixed	Eb	Egbungi series, brown	As Egbungi above, but with sandy subsoil lenses or shallow clay (less than 1 m) over sand; poorly and imperfectly drained.						
	Br	Brung series	Up to 60 cm of mottled brown hard concretionary loamy sand over grey mottled clay and sandy clay; imperfectly drained. May be sodic.		8.1* 8.4*		8		ts ss

a = % clay
b = pH, KCl; * = water
c = % organic matter
d = cation exchange capacity, meq/100g
e = % base saturation
f = stratigraphy; ts = topsoil (0-20 cm), ss = subsoil (40-120 cm)

Table 1. Description of soil mapping units

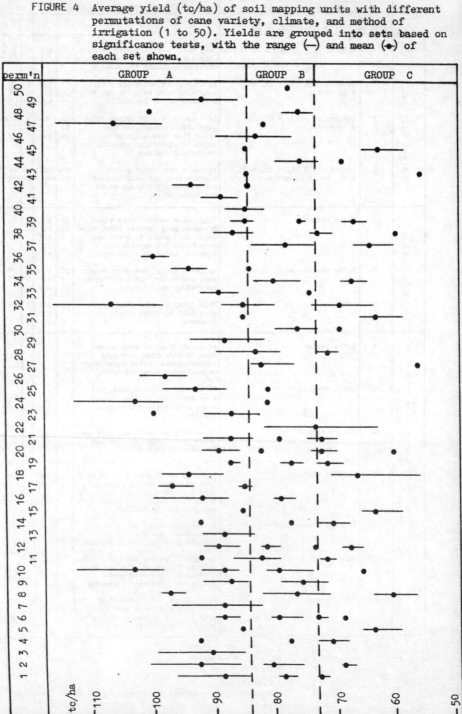

FIGURE 4 Average yield (tc/ha) of soil mapping units with different
permutations of cane variety, climate, and method of
irrigation (1 to 50). Yields are grouped into sets based on
significance tests, with the range (⊢⊣) and mean (•) of
each set shown.

FIGURE 5

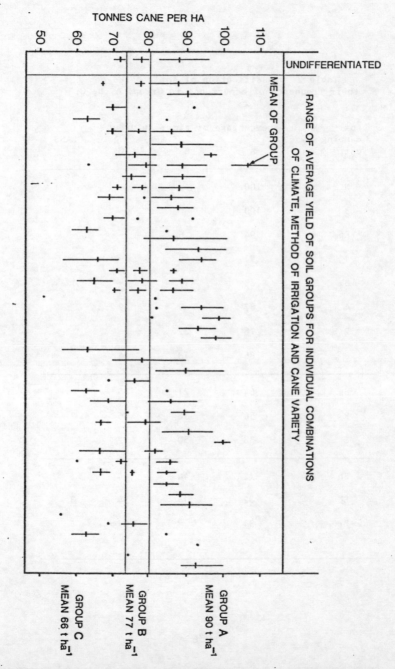

Table 2 Classification of soils based on
their frequency of occurrence in Groups A, B, C.

soil mapping unit	percentage of times in group A	B	C	CLASS
	A	B	C	
N	100	–	–	
3	100	–	–	
Ep	94	6	–	
1	91	9	–	
B	83	17	–	1
4	82	9	9	
2	78	22	–	
0	86	10	4	
Bs	64	21	15	
R	56	38	6	2
E	38	50	12	
Sh	12	44	44	
Eb	32	14	54	3
Br	31	8	61	

Table 3   Relative suitability of soils for irrigated estate
sugar-cane, from existing references (A to E)

A	B	C	D	E
Very good 1,2,E,Ep	Very good O,E,B			
Good, Fair 3,4*,R	Good	Suitable 1,2,3,E,Ep	Definitely suitable 1,2,O	Most suitable E,O,B
Poor N,O,B,Bs,Sh	Poor N	Moderately suitable O,B,Bs N,R,Sh	Probably suitable 3*,4* Bs,Sh	Poor Br
Very poor Eb,Br	Very poor Br,Sh			Least suitable R,Sh,N
Unsuitable 4*		Unsuitable 4,Br,Eb	Unsuitable 3*,4*	

*Depending on complexity of soils in mapping units

Sources: A Land Resources Development Centre (1972)
          B Innes (1965)
          C NEDECO (1961)
          D Smith (1972)
          E Higgins et al (1960)

their potential, with significant differences between mapping units. With overhead irrigation, where water application and disposal are better controlled, all the mapping units yield similarly (no significant difference between them) and at about 80% of potential.

CONCLUSIONS: A WAY AHEAD

There is no reason to suppose that Bacita is atypical of tropical irrigation schemes, so the conclusions may be applicable elsewhere.

In the first place, knowledge of the soils is important for identification of potential limitations and development of management guidelines, but apparently large and obvious differences between mapping units may not always be significant when trying to assess relative suitability.

Second, whereas the effect of one or two parameters on yield may be known, predictions of the effect of many environmental factors in combination is still subjective. Because of this, any suitability assessment is also subjective. So why do soil surveyors persist with land suitability evaluation?

Results at Bacita have shown that original assessments were wrong in several respects. It seems that there is little justification for attempting to define degrees of suitability once an area has been identified as broadly suited to the specified use. Certainly in some circumstances some soils will yield consistently well while others will do poorly, but in other circumstances the positions may be reversed or equalised. Too little is known about the circumstances or their effects to make any worthwhile estimate of relative soil suitability. Although it is considered that the optimum soil for cane is a well-drained clay loam, the best soil at Bacita (Nebung) is imperfectly-drained sand over loamy sand!

How then can the speculative interpretation of land suitability be avoided, and at the same time tell the user what he wants to know? We have defined the suitability of a mapping unit for a specified use as an assessment of yield relative to cost of obtaining that yield. Both yield and cost for each mapping unit will vary. Neither relative nor absolute yield from mapping units can be predicted accurately when management capability is relatively low, as at Bacita and elsewhere in tropical Africa. The cost of developing and maintaining production of each mapping unit can be ascertained, although usually at the development stage of a project, rather than at the feasibility stage.

The need for a suitability ranking of mapping units is questionable. It is therefore unnecessary, and vanity on his part, for the soil surveyor to persist with "ritual" land suitability evaluation, other than to state whether or not the land can be used as intended. Some production estimate must be given, but it is more realistic to give an overall sustainable average yield, bearing in mind the future management capability. This can be gauged from yields presently being obtained from Government and University Research Stations in the country or in comparable environments. Then, by ascertaining actual yields from other farms and projects, a current "management factor" (actual as a percentage of potential yield) for the country can be deduced: at Bacita this is 70%.

By assuming one realistic overall average yield, land suitability assessment can then be based on economic criteria. Knowing the inputs necessary for any particular mapping unit (based on the soil survey data), the cost of developing and continuing to cultivate that mapping unit as opposed to others can be evaluated.

REFERENCES

Aitken, J.F. (1983) Relationships between yield of sugar cane and soil mapping units, and the implications for land classification. Soil Survey and Land Evaluation 3, (1): 1-9

Burgess, T.M. and R. Webster (1980) Optimal interpolation and isarithmic mapping of soil properties.
    I. The semi-variogram and punctual kriging
    II. Block kriging J. Soil Science 31, (1): 315-342

FAO (1976) A Framework for Land Evaluation (edited by R. Brinkman and A. Young)
FAO Soils Bull. 32, Rome.

Higgins, G.M., D.M. Ramsay and D.L. Story (1960) Report on the second soil survey of Bacita Fadama, Ilorin province, N. Nigeria. Soil Survey Bull. 8, Ministry of Agriculture, Samaru, Zaria, Nigeria.

Innes, R.G. (1965) unpublished report.

Land Resources Development Centre (1972) unpublished. Overseas Development Administration, Surbiton, UK.

NEDECO (1961) Niger Dams Project vol. 5, 7 Irrigation and Agriculture, report to the Federal Government of Nigeria.

Northcote, K. (1984) Soil-landscapes, taxonomic units and soil profiles. Soil Survey and Land Evaluation 4 (1): 1-7.

Smith, W. (1972) Feasibility Study and Development Proposals, Phase II Development. Booker Agricultural and Technical Services Limited and Bacita Estate, confidential report.

Taylor, R.D. unpublished soil map of Bacita Estate.

ACKNOWLEDGEMENTS We would like to thank Booker Agriculture International Ltd. for their cooperation in this research, and the Nigerian Sugar Company Ltd (Bacita Estate) for their help in providing access to data.

IRRIGATION MANPOWER PLANNING WITH SPECIAL REFERENCE TO NIGERIA

R.C. Carter, M.K.V. Carr, M.G. Kay, T.L. Wright

INTRODUCTION

Nigeria is unique in Africa in having not only extremely ambitious
plans for irrigation development, but also the financial resources to
implement her intentions. (In the current five year plan, 1981-85,
around 1.4 million ha are identified for irrigation development, and
planned expenditure over that period is estimated at about 2 billion
Naira). However it has been widely recognised that lack of suitable
manpower is likely to be a serious constraint to the achievement of
such development plans (Nwa, 1982; Patil, 1980; Oyaide, 1981; Aliyu
and Kaul, 1980; Onazi and Patil, 1979).

It was therefore timely that the authors were invited to carry out
a countrywide manpower planning study for the irrigation sector
recently (Wright et al, 1982). The objectives of this project,
undertaken on behalf of the Nigerian Government, were to assess the
manpower needs and training requirements for irrigation over the next
twenty years or more, and to set out a strategy for their realisation.

This paper briefly describes the methodology adopted in the study
and highlights the main difficulties of irrigation manpower planning
in a country which has only a very short history of formal irrigation
development.

Because of the paucity of published work on irrigation manpower
(especially in regard to Africa), much of the present study had to be
carried out from first principles. The planning exercise was
approached as a supply and demand situation. Figure 6 shows
diagrammatically the approach taken first of all in evaluating the
problem, and then in arriving at a strategy for meeting Nigeria's
irrigation manpower needs.

In the course of the study it was necessary to visit and evaluate
as full and representative as possible a range of the following
activities and institutions in the irrigation sector:

    educational institutions (at all levels)
    training bodies research organisations
    government departments (at both Federal and State level)
    formal irrigation projects
    traditional irrigation activities
    consultants and contractors

IRRIGATION AND IRRIGATION MANPOWER

Two related questions immediately arise at the beginning of a study
of this type: the first concerns the scope of and need for irrigation
in a country as diverse as Nigeria. The climate varies markedly from
south to north, so that for example rainfall in the extreme south east
reaches 4 000 mm per year, spread over 11 months, while in the far
north east annual rainfall is less than 500 mm and concentrated in
three to four months. One important consequence of the variation just

FIGURE 6 Manpower Planning Methodology

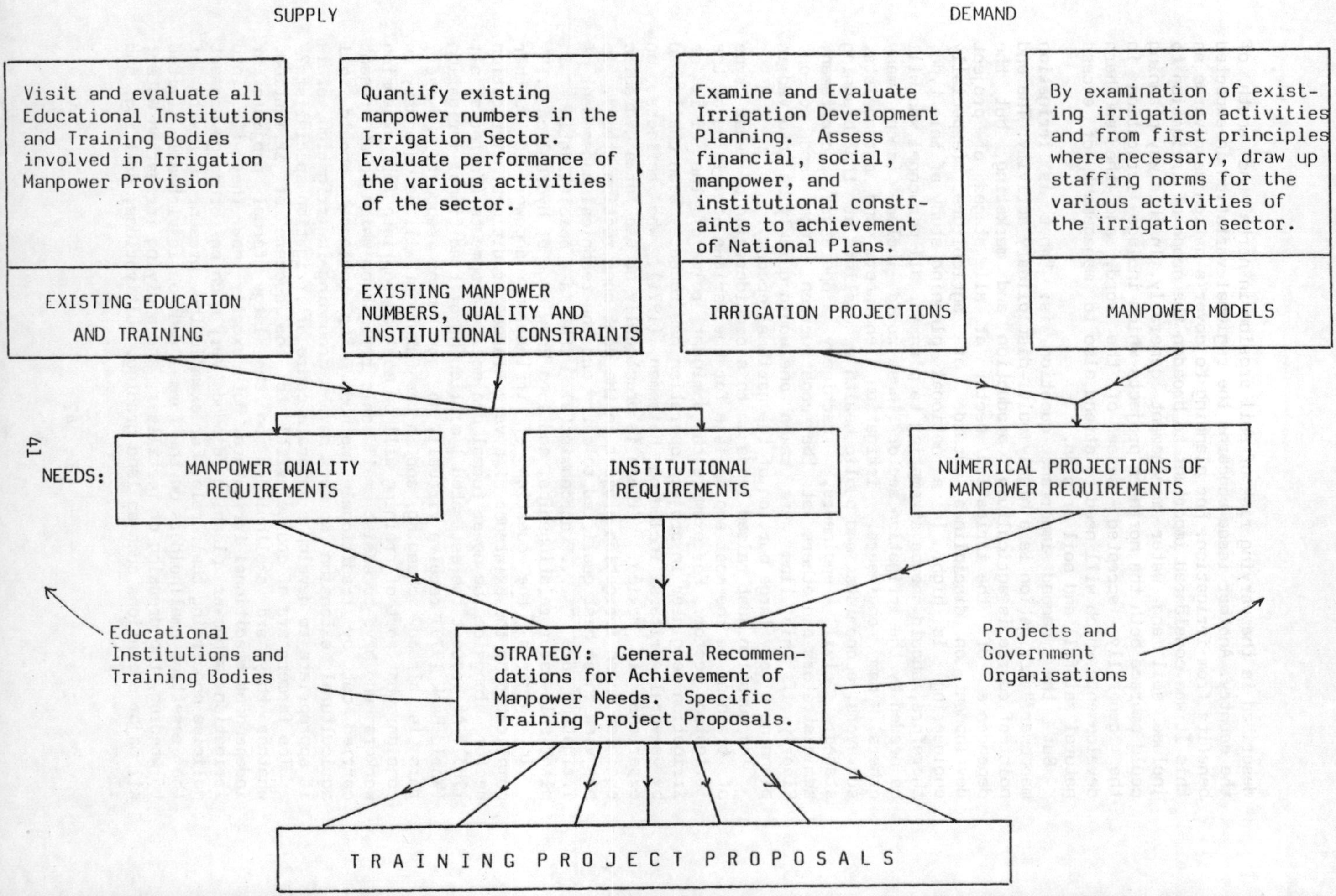

SUPPLY

DEMAND

Visit and evaluate all Educational Institutions and Training Bodies involved in Irrigation Manpower Provision.

EXISTING EDUCATION AND TRAINING

Quantify existing manpower numbers in the Irrigation Sector. Evaluate performance of the various activities of the sector.

EXISTING MANPOWER NUMBERS, QUALITY AND INSTITUTIONAL CONSTRAINTS

Examine and Evaluate Irrigation Development Planning. Assess financial, social, manpower, and institutional constraints to achievement of National Plans.

IRRIGATION PROJECTIONS

By examination of existing irrigation activities and from first principles, where necessary, draw up staffing norms for the various activities of the irrigation sector.

MANPOWER MODELS

NUMERICAL PROJECTIONS OF MANPOWER REQUIREMENTS

Projects and Government Organisations

NEEDS:

MANPOWER QUALITY REQUIREMENTS

INSTITUTIONAL REQUIREMENTS

STRATEGY: General Recommendations for Achievement of Manpower Needs. Specific Training Project Proposals.

Educational Institutions and Training Bodies

TRAINING PROJECT PROPOSALS

described is the varying risk of soil erosion in north and south of the country. Another issue concerns the regional variation in expected benefits from irrigation. For planning to cover a nation as diverse as this it was considered important to broaden the scope of irrigation to include soil and water management generally. In this way planning could embrace both the northern projects which include irrigation in the generally accepted sense of the word, and the southern developments which will need to attend also to management of excess natural rainfall and soil erosion.

But the second important question is: "What is irrigation manpower?" Irrigation is a highly multi-disciplinary activity. The end point of course is agricultural production and marketing, but the dependence of the irrigation sector at all stages of project development on disciplines outside of agriculture (especially engineering) is high. In a national planning study of this type, therefore, should one be attempting to plan for all personnel who will be needed by the irrigation sector? These would include (among many others) dam engineers, irrigation engineers and agronomists, surveyors, economists, and public health specialists at the planning stages; civil engineers, machinery operators, craftsmen, administrators and others at the construction stage, and so on. Clearly if this line were taken one would quickly be involved in planning a very large portion of the national economy.

The question just raised leads on to a consideration of what type of training is the most appropriate for the various activities of the irrigation sector. For example the dominant role of the civil or irrigation engineer in project operation has recently been challenged by several writers including Haissman (1971), who reflects on experience in Mexico: "there is practically no use for a hydraulic engineer on an irrigation project in the post construction stage". In his view those best qualified to carry out the technical management of irrigation are " ... agronomist(s) with a specialisation in plant-soil-water relationships, and some knowledge of hydraulics ..."

In Nigeria the question of irrigation manpower is further complicated by the existence of a very large traditional irrigation sector alongside the newer formal project developments (Carter et al, 1983). At present the estimated 1 million ha of traditional irrigation (World Bank 1979) receive virtually no government support. However, this is already changing, and in the future it will be necessary to consider what sort of training will be most appropriate for extension workers who have to assist small scale irrigation which is at present carried out by traditional methods. For instance should all agricultural extension workers have a grounding in irrigation, or is it appropriate to develop a separate cadre of irrigation specialists?

The farmers are of course vital to the success of irrigation, whether they are smallholders on the large formal projects, or independent traditional irrigators. But exactly how they fit into irrigation manpower planning depends very much on future government policies concerning their role. For example the present activity of the resettled smallholders on the formal schemes is virtually limited to weeding, top dressing of fertiliser, and applying irrigation water; all major operations such as land grading, cultivations, drilling, and

42

harvesting are meant to be carried out by the project organisation. If the role of the farmers is to expand then the training needs of both farmers and project staff will alter, depending on the evolving relation between these two groups.

In the present study irrigation manpower was taken to include the following:

1) Personnel at all levels including farmers trained or needing to be trained in the core irrigation specialisms, namely irrigation agronomy, soil and water management, and irrigation engineering.

2) Those of other disciplines or skills whose expertise is severely lacking at present and therefore a major constraint to the success of irrigation development. In Nigeria it was considered that the most important general skills which are missing are those of management.

IRRIGATION MANPOWER DEMAND

In assessing manpower demand for the future an attempt must be made to forecast irrigation development nationally over the period of the plan. Forecasts of this nature are fraught with difficulties, as is illustrated by previous unfulfilled projections of performance in the Nigerian irrigation sector (Elahi, 1971; World Bank, 1979, see Figure 7)

Any such forecasting must take into account not only the available National Plans, but also an assessment of the constraining effects of financial, social, manpower, and organisational limitations to development. All this may be somewhat subjective, although it can be better informed by the fullest consultation with professionals in the field.

However projections can at least be given an air of realism by a consideration of lead times in project development. In Nigeria, where many planned formal irrigation projects or project phases are in the order of 20,000 ha, a phasing along the following lines is considered to be possible (barring major constraints of the types listed above)

Phase	Years
Regional studies, Feasibility studies, Designs, preparations of Contract Documents	1-3
Dam Construction	4-8
Construction of Irrigation and Drainage System	8-11
Land preparation and Resettlement	9-13
Full Operation	14 onwards

This sort of scheme is of course very generalised, but for planning purposes it can at least put an upper bound on the forecast rate of developments.

The second element in the assessment of future manpower demand involves a knowledge of appropriate staffing levels for the different parts of the irrigation sector. (For this and other parts of the

FIGURE 7

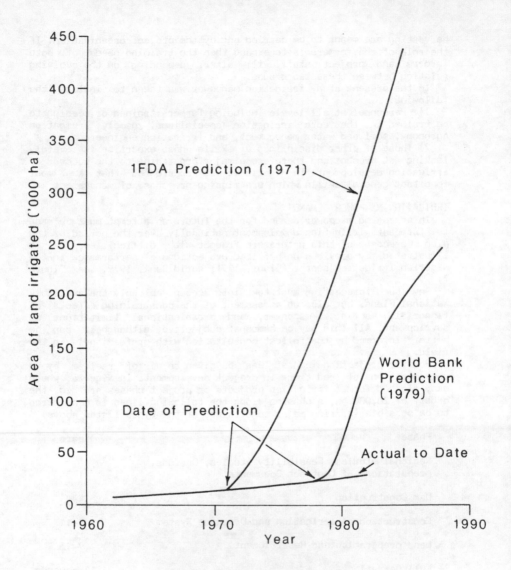

1FDA Prediction (1971)

World Bank
Prediction
(1979)

Date of Prediction

Actual to Date

exercise, the irrigation sector was broken down into five sub-sections, namely, planning and design, construction, operation and maintenance, research, and education and training. The organisations involved in these sub-sectors include Government departments, consultants, contractors, educational institutions, and research bodies). Very little relevant published information is available in this area, and often existing practice cannot give the appropriate data. Almost the only published work at the time of the present study was that of Haissman (1971), referring to Mexico, and that of Bottrall (1981) discussing Asia and the Far East. Since the publication of our report, the FAO has brought out a valuable paper which includes useful generalised figures for project operation, maintenance, and administration staffing levels (Sagardoy, 1982). Our own estimates of staffing norms were based on the previous experience of the team with Nigerian irrigation, and on comparison with existing staffing levels in the country and elsewhere.

The third part of the demand assessment consists of the fusing of the irrigation projections with the staffing norms to produce numerical manpower requirements. In doing this full account has to be taken of wastage, education lead times, and expected productivity increases. Estimation of wastage was difficult in Nigeria where significant numbers of trained personnel opt out of employment in poorly paid positions on government projects which are often in remote places.

PRESENT SUPPLY OF IRRIGATION MANPOWER

Having defined irrigation manpower for the purposes of the planning exercise (above) it is necessary to visit and assess as many relevant educational institutions and training programmes as possible. Appropriate university level education may take place in Departments or Faculties of Agriculture, Civil Engineering, or Agricultural Engineering. In the Nigerian study most attention was paid to the departments of Agricultural Engineering or their equivalent, since irrigation formed a very minor component of courses in other departments. At sub-degree and vocational level in Nigeria a wide range of appropriate courses in irrigation agronomy, irrigation engineering, agricultural mechanisation, and soil conservation, among others, are training irrigation manpower.

Training programmes outside of educational institutions can be a second important source of irrigation manpower. These may be on the job training schemes within irrigation projects, or local and regional training programmes for and by extension workers.

It is desirable to make an assessment of education and training facilities in subject areas which may not strictly or only relate to irrigation, but which nevertheless may form constraints to its success. In Nigeria such a field is that of management, so special attention was paid in the study to the available training resources for those skills.

The assessment of the education and training facilities outlined in the preceding three paragraphs has a number of components. Broadly it consists of the following:

(i) Estimation of present and planned student numbers.

45

(ii) Examination of resources such as laboratories, workshops, field sites, and equipment, as well as teaching facilities, textbooks etc.

(iii) Examination of curricula.

(iv) Estimation of teaching staff numbers.

(v) Assessment of constraints to achievement of present and future objectives.

Finally in the supply side of the planning exercise it is necessary to quantify existing personnel numbers in the various activities of the irrigation sector. Estimates of manpower numbers, both indigenous and expatriate, in the activities and organisations mentioned above are necessary to evaluate the shortfall between demand and supply.

CONSTRAINTS TO MANPOWER PERFORMANCE

So far in this paper the emphasis has been largely on manpower numbers. However even given sufficient numbers two factors may militate against the effective performance of the available personnel. Firstly as individuals they may lack important skills or abilities, a problem which can conceivably be overcome by more or better training. And secondly the organisations in which they work may fail to provide adequate incentives, supervision, and rewards, so that the full potential of the staff is not realised.

The first of these factors can be referred to as the issue of manpower "quality". Aspects of this problem in many developing countries include a widespread lack of practical skills and an unwillingness to be involved in practical tasks, as well as poor management abilities. Both these areas can be dealt with by appropriate training. In particular they could both be improved through better use of on-the-job training periods which, in Nigeria at least, are amply allowed for in educational timetables but very poorly supervised at present. These issues of manpower "quality" must be dealt with at source; it is no solution to the manpower problem to overstaff with poorly trained or incompetent personnel.

Shortcomings in the working environment relate to three main aspects:

(i) Poor organisational structures, which discourage initiative and which fail to provide career incentives;

(ii) Inadequate management systems which do not set objectives of targets and work towards their achievement;

(iii) Lack of coordination and communication between related government organisations.

All these institutional issues are important aspects of manpower planning since it is of limited value to train personnel when the environment in which they have to work is far from ideal. Bottrall (1981) also makes this point.

SETTING OUT A STRATEGY FOR THE FUTURE

The 'bottom line' of any manpower plan is the strategy which it sets out for realising the manpower needs. A number of important points can be made about this strategy.

(i) As has been indicated already the plan must take account of the required numbers of workers, the content of their training, and also the management of the organisations in which they work. It will therefore focus not only on the educational institutions and training bodies, but also on the irrigation projects and government organisations (see Figure 6).

(ii) In order to remain realistic and relevant the plan must propose some means for its own updating. One approach to this is to adopt a rolling Plan system so that at all times the best knowledge and estimates of demand and supply keep the plan fresh. The implementation of this procedure necessitates a standing committee or other body to take responsibility for 'rolling' the plan.

(iii) Finally it is important to express the strategy not only as general recommendations but also as specific proposals for action. In the present study ten manpower development projects were outlined, each of which would contribute to the overall strategy for meeting the national irrigation manpower needs. These varied widely in scope and cost, so that possible funding could be attracted from a variety of different agencies.

CONCLUSIONS

This paper has raised some of the problems encountered in a manpower planning study of the irrigation sector in Nigeria. It remains to summarise the main conclusions for irrigation manpower planning in general.

(i) There is a danger in thinking too narrowly about irrigation as an activity, and the types of training appropriate to irrigation personnel. It may be necessary to go back to first principles in defining the activities of the irrigation sector which are most appropriate to the natural environment, and also the corresponding curricular needs in education and training.

(ii) The difficulty of forecasting development of any kind is considerable. Financial, socio-political, environmental, and manpower constraints all can drastically affect national plans. For planning purposes though it is necessary to be somewhat optimistic in making future projects, to avoid the possibility of manpower remaining a major constraint to progress. The question still remains however as to how projections can be made most realistically and whether a Rolling Plan approach can significantly help.

(iii) Staffing norms ideally should be derived from an examination of well managed irrigation in the country in question. However if there is no such suitable basis for arriving at the figures, alternative means have to be found. The published literature is sparse on this subject, so further work is clearly needed in this area.

(iv) On the supply side of the problem there are difficulties in quantifying wastage, especially by movement from poorly paid government employment into the private sector. In Nigeria specifically the remoteness of several of the irrigation projects and educational establishments has led to problems of recruitment, and the proliferation of new institutions has led to great shortages of staff for all activities of the irrigation sector. These problems all need to be taken into account in the planning.

(v) The widespread need for practical and management skills in the irrigation sector raises questions as to the appropriateness of conventional educational qualifications. Certainly far more effort and resources should be put into practical training during educational courses, and perhaps a greater emphasis should be put on operation and management rather than design. But there is also a need for effective systems of 'sandwich year' training, counterpart staffing, and on the job training at all levels.

(iv) The common institutional shortcomings outlined above point to the need for reform of organisational structures and management systems. This is a very sensitive political area and the intentions behind any such proposals may be misconstrued. However an adequate manpower plan must pay attention to this area.

(vii) Finally irrigation manpower planning has to take account of farmers' roles in both formal and traditional irrigation. The system of extension to traditional irrigation farmers, and the relationships between farmers and project staff on the government schemes will affect both required manpower numbers and the types of training needed at all levels.

REFERENCES

Aliyu, M.B. and J.N. Kaul (1980). Lack of adequate education and training for junior cadre manpower as an alarming impediment component in River Basin Development Authorities: A case study in respect of RBDAs. School of Irrigation and Extension Services. 7th National Irrigation Seminar, Bagauda Lake, Kano State, September 8-12, 1980.

Bottrall, A.F. (1981) Comparative study of the management and organisation of irrigation projects. World Bank Staff Working Paper No. 458, World Bank, Washington, D.C.

Wright T.L., R.C. Carter, M.K.V. Carr, G.K. Kay (1982) Nigeria: Manpower Needs and Training Requirements for Irrigation (MANTRI). British Council, London.

Carter, R.C., M.K.V. Carr and M.G. Kay (1983) Policies and Prospects in Nigerian Irrigation. Outlook on Agriculture 12, (3): 75 -76.

Elahi, A.M., J.A. Raza, and Z.G. Tyson (1971) National Agricultural Development Committee: Report of the Study Group on Irrigation and Drainage. Federal Department of Agriculture, Lagos.

Haissman, I. (1971) Generating skilled manpower for irrigation projects in developing countries: A Study of Northwest Mexico. Water Resources Research 7 (1): 1-17d.

Nwa, E.U. (1982) Development and progress of agricultural engineering in Nigerian Universities. Transactions of the American Society of Agricultural Engineers, 25, (2)

Onazi, O.C. and S.S. Patil (1979) Manpower requirements and

training in Irrigation Schemes in Nigeria. Sixth Nigerian Irrigation Seminar, Zaria, September 26-28, 1979.

Oyaide, O.F.J. (1981) Agricultural manpower development in Nigeria: The Federal department of Rural Development and the Integrated Rural Development Project Management Training Institute (ARMTI), Badagry, July 22-24, 1981.

Patil, S.S. (1980) Manpower requirements and training in irrigation schemes in Nigeria. Paper for Sixth Nigerian Irrigation Seminar, Zaria, September 26-28, 1979.

Sagardoy J.A. (1982) Organisation operation and maintenance of irrigation schemes. FAO Irrigation and Drainage Paper 40, FAO Rome.

World Bank (1979) Nigeria Agriculture Sector Review. World Bank, Washington.

FUELWOOD: A FORGOTTEN DIMENSION OF IRRIGATION PLANNING

Francine M.R. Hughes

INTRODUCTION

This paper will consider the general issue of fuelwood needs in poor countries and more specifically, fuelwood needs on irrigation schemes. The problems encountered in providing fuelwood for the Bura Irrigation Settlement Project in Kenya and the lessons learnt from this case study will be discussed in more detail.

THE FUELWOOD DILEMMA

About nine-tenths of the people in most poor countries today depend on firewood as their chief source of fuel (Eckholm, 1975). The continent with the highest per caput consumption of firewood is Africa (Table 4). In most poor countries, the growth of human populations is outpacing the growth of new trees and as Eckholm (1975) puts it, this is leading to a "costly diversion of animal manures to cooking food rather than producing it and an ecologically disastrous spread of treeless landscapes." The scale of fuelwood scarcity is discussed in several places in the available literature, for example, Kamweti (1981) and Shakow et al (1981). They point out that at one end of the scale, when wood is plentiful, people consume it wastefully but only use dead wood and when wood becomes scarcer, branches are cut from live trees and whole trees are felled. At the other end of the scale, people use agricultural waste, sawdust and cowdung and eventually change their eating habits so that food items like beans, which take a long time to cook, are removed from the diet. Hoskins (1980) also comments on the adverse change of diet as a result of fuelwood shortages. In a study carried out in general, she found that as a result of decreased fuelwood availability, women had changed from serving two hot meals per day to one and eventually to one every other day. On the days in between hot meals, the women served their families uncooked millet flour mixed with water because there was no fuel with which to cook.

Various alternative sources of energy have been considered for many rural areas. Broadly speaking, these are either commercial, non-renewable forms of energy or the substitution of alternative renewable energy sources such as biogas and solar power. The first alternative is usually too expensive, especially in terms of foreign exchange and the second altrenative can only be of help to a minor extent (Openshaw, 1981). The only lasting solution, therefore, to providing enough wood for rural people to cook their food is to increase tree planting.

CREATING NEW FUELWOOD RESOURCES

Three scales of tree planting programmes are usually considered:
1) Large-scale plantations;
2) Community or village woodlots;
3) Farm or individual planting.

Table 4   Fuelwood in World Energy Consumption in 1978

	Population (millions)	Total Fuelwood[1] (millions of m³)	Consumption per capita (m³)	Energy Equivalent of Fuelwood[2] (millions of gigajoules)	Commercial Energy[3] (millions of gigajoules)	Fuelwood (percentage of total)[4] %
World	4 258	1 566	0.37	14 720	256 594	5.4
Developed world	1 147	145	0.13	1 363	205 115	0.7
Market economies	775	54	0.07	508	145 148	0.3
Centrally planned economies	372	91	0.24	855	59 967	1.4
Developing World	3 111	1 421	0.46	13 357	51 479	20.6
Africa	415	353	0.85	3 318	2 415	57.9
of which least developed countries	138	163	1.18	1 532	255	85.7
Asia	2 347	796	0.34	7 478	37 558	16.6
of which least developed countries	130	34	0.26	319	180	63.9
Centrally planned economies	1 010	220	0.22	2 068	24 048	7.9
Latin America	349	272	0.78	2 557	11 306	18.4
of which least developed countries						

Notes: 1 Includes wood for charcoal
2 1 m³ = 9.4 gigajoules
3 INT coal = 29.3 gigajoules
4 Not including other sources of non-commercial
   energy important in some regions.

Source: Report of the Technical Panel on
Fuelwood and Charcoal to the United
Nations conference on New and Renewable
Sources of Energy, Nairobi, 1981)

It is stated in the report of the 'Technical Panel on Fuelwood and Charcoal' to the United Nations conference on New and Renewable Sources of Energy (UNERG) held in 1981, that "the provision of new fuelwood plantations presents not serious technological problems but rather institutional, social and economic problems apart from the possible competition for suitable land for other uses." The report further states that "fuelwood supplies should be established not only in the form of fuelwood plantations, from individual to large scale but also as an integral component in agriculture and rural development programmes." (UN 1981) Above all emphasis is placed in the literature, on the involvement of the local people in any new planting efforts, for example, Hoskins (1980), Bhatt (1981), Kamweti (1981) and Openshaw (1981). Thus, the UN (1981) report comments, "both governments and international organisations should make special efforts to involve women in actively participating in fuelwood programmes, particularly in aspects related to their usual responsibilities of supply and use" and again "for rural plantation programmes specific needs and preferences of the users need careful consideration".

FUELWOOD NEEDS ON IRRIGATION SCHEMES

The widespread problem of inadequate fuelwood supplies for rural populations becomes concentrated in association with irrigation schemes. Whether or not an irrigation scheme is built to resettle people who already live in the area or to resettle people who come from elsewhere, provision of firewood for farmers on an irrigation scheme directly or indirectly poses a problem for the irrigation scheme managers. This is largely because construction of irrigation schemes clears many areas of bush and trees which would normally be a source of fuelwood for local people. The design of irrigation schemes with their systems of canals and large fields inevitably means that women on the schemes have to walk considerable distances to reach suitable areas of bush for firewood collecton.

If, as in the case of the Gezira Scheme in the Sudan or the Bura Irrigation Resettlement Project in Kenya, the scheme design incorporates a fuelwood plantation, then the manager has assumed direct responsibility for supply of wood to the tenant farmers and taken on the considerable task of successfully implementing a fuelwood plantation. However, implementation of an irrigated plantation should be relatively easy at the inception of a new irrigation scheme as the plantation is simply an extension of the area being prepared for crop irrigation, assuming that land is available. In the case of the scheme at Bura, the firewood plantation was included in the design because the scheme's primary aim is to resettle people from more crowded parts of Kenya into less populated areas and fuelwood has to be supplied to offset the sudden, massive local population increase. Fortunately for the Bura scheme managers, land is available for an irrigated plantation.

If, as in the case of the Bakolori Project in Nigeria, fuelwood plantations have not been set up as part of the irrigated area, then the scheme manager has assumed no responsibility for fuelwood supply and may have acquired the problem of losing farmers from the scheme because fuelwood collection has become too difficult. At Bakolori, no provision was made for fuelwood because the farmers on the scheme already lived in the area and the scheme merely provided agricultural services to an already existing farming population. The fact that scheme construction involved considerable bush and tree clearance was not taken into account. In this sort of situation, later provision of fuelwood from plantations is much harder, as a second construction phase is needed and anyway, the lack of space for such a plantation in an area that is already densely populated poses a major problem

This paper will now consider in detail the problems encountered on the scheme at Bura where, seemingly, provision of firewood should have been relatively easy. It is hoped that consideration of these problems will demonstrate how difficult it is to provide enough firewood for an irrigation scheme and point to the necessity for much more serious planning of firewood provision in the early stages of a project.

THE IRRIGATION SCHEME AT BURA

The Bura Irrigation Settlement Project is being constructed alongside the Lower Tana River of Kenya (Figures 8 and 9). The whole scheme may eventually comprise three parts, a large area on the East bank which could cover as much as 26,000 hectares, though it now seems unlikely that it will ever be built, and two smaller areas on the West Bank, the second of which will incorporate the pilot scheme of 25 years standing, at Hola. The first of the West bank schemes (called Phase I) is now partially irrigated and will cover 6,700 ha when completed. If the second area goes ahead, the total irrigated area on the West bank, including the 800 ha at Hola will be approximately 12,000 ha. The whole scheme has been designed for growing cash cotton to increase foreign exchange earnings and for resettling people from crowded parts of Kenya.

The area in which the scheme is built is semi-arid with dry scrub vegetation except in the immediate vicinity of the river where a ground-water fed evergreen forest extends for about one kilometre on either side of the river and for about 300 km of river length. The irrigation scheme structures parallel approximately 80 km of the riverine belt. The local population consists of three tribes. The Malekote and Pokomo are sedentary agriculturalists living along the river banks and number about 15,200 in the area to be influenced by the irrigation scheme. The Orma are semi-nomadic pastoralists who number about 6,400 in the same area. Phase I of the West Bank Scheme will accommodate 5,150 tenant farmers plus their families in 23 scheme villages and it is estimated that this part of the scheme will eventually support approximately 60,000 people between the farming families and the commercial and administrative sectors. The increased pressure on local fuelwood supplies will therefore be at least four-fold.

PROVISION OF FUELWOOD ON THE BURA SCHEME

Provision of fuelwood for tenant farmers has been seen as a critical problem for the irrigation scheme since the planning stage and several alternative sources of fuelwood were considered. It was eventually decided that the least-cost method of meeting the project's

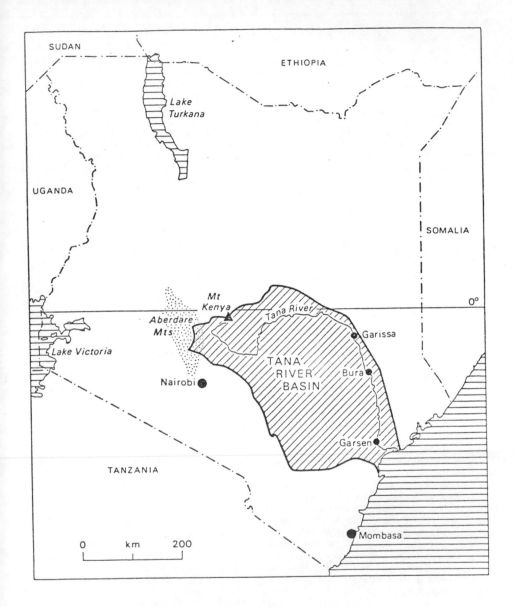

FIGURE 8   Location of Tana River Basin

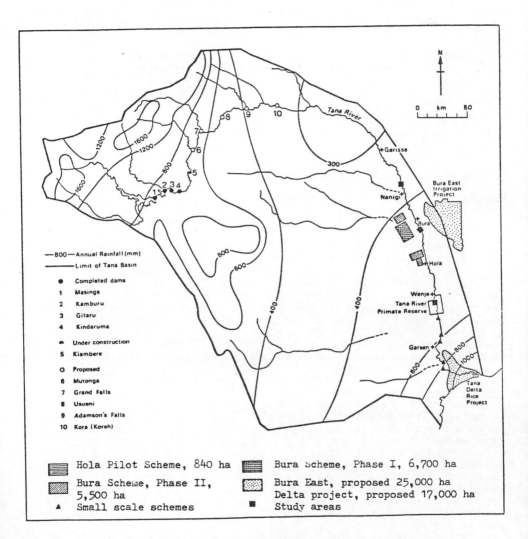

FIGURE 9   Location of development schemes in the Tana
Basin and annual rainfall

demands was to use gravity-fed irrigated plantations adjacent to the project area with additional dead wood being taken from the riverine forests on a controlled basis. (See Figure 8) Gazettement of the forests for conservation reasons precludes further use than this. The need for alternative sources of fuelwood has been highlighted by the experience on the Hola Scheme where, during the 25 years of project operation, the dry bush has been devastated over increasingly large areas and the local riverine forest has almost been destroyed. At present, firewood is collected from considerable distances and transported to the town of Hola by bicycle. In the Hola Irrigation Scheme villages, the women still walk to collect their wood but now spend over half a day each time they do so.

PLANNING THE IRRIGATION PLANTATION AT BURA

The exact area that needs to be under irrigated forestry is clearly a function of demand for wood and plantation productivity. Ideally, a survey of firewood consumption should have been carried out on the Hola Scheme, to give some idea of the needs of the Bura Scheme and therefore, to assist decisions on plantation size at Bura. The Project Planning Report (PPR, 1977) bases its estimates on the figures given in a World Bank report which estimates Kenyan figures of fuelwood consumption to be approximately 2 kg per person per day (using an average specific gravity factor of 0.75, this is 0.991 cubic metres per person per year (PPR, 1977). The figure eventually used for the irrigated plantation at Bura was 0.7 m³ per person per year.

Forest species trials have been carried out at Hola since 1965 but almost no usable results were documented until the start of a new set of trials in 1979 to choose species suitable for the plantations at Bura and to give some idea of how fast trees could grow under different irrigation regimes. Thus, when the plantations were designed, the PPR (1977) assumed an average mean increment of 15 cubic metres per hectare per year which was reckoned to be a conservative estimate. A total of 4,500 ha was eventually arrived at as the size of irrigated plantation needed for Phase I of the Bura Scheme alongside the 6,700 hectares of cropped area.

Construction of the irrigation scheme began in 1973 and the first tenant farmers moved onto the scheme in 1981. By the time they arrived to plant the first crop, no tree seedlings had been sown in the areas put aside for the plantations, despite it being obvious to the scheme managers, the aid agencies funding the scheme, and all people involved in scheme design and construction that unless trees were planted at the very beginning of the construction phase, no trees would be ready to harvest for firewood when the farmers arrived. An attempt to alleviate the problem was made by stock piling the bush cleared off irrigated areas for villagers to use as firewood, but this has by no means been adequate.

Although awareness of the firewood need was present from an early stage in the scheme plans, several factors have contributed to unsuccessful implementation of the irrigated plantations. These include:

1) No clear idea on the exact size of plantation needed and therefore a lack of incentive to implement it.

2) Lack of specific and adequate funds for setting up of the plantation, notably nursery facilities and extra irrigation infrastructure.

3) Poor monitoring of project progress by the aid agencies funding the scheme and therefore, inadequate pressure by them on the scheme manager (The National Irrigation Board of Kenya) to ensure sufficient firewood supplies for farmers already resident before encouraging more farmers to move on to the scheme.

4) Poor communication between the government departments involved in setting up the irrigated plantation, notably, the National Irrigation Board and the Forest Department of the Ministry of Environment and Natural Resources.

The author attempted to collect data on levels of firewood consumption in the Lower Tana during 1981 and 1982 to try and fill in some of the information gaps that have caused problems with planning and implementing the irrigated firewood plantations at Bura.

DATA COLLECTION

In the present study, an initial survey was carried out of levels of firewood use in three Hola Scheme villages, two Bura Scheme villages and in villages along a selected section of river. At Bura, the two villages that were first settled were studied (Villages 1 and 2) as other villages were not considered to have established regular patterns of firewood collection at the time of study. In all cases, the women of selected households were interviewed to discover the frequency of firewood collection, size of household, number of hearths and meals cooked. Many were accompanied in their wood forays so that distance travelled and species used could be noted. Each woman's bundle was weighed upon return to her village. A total of 39 Hola Scheme households, 40 Bura Scheme households and 33 riverine households were studied in this way. The whole survey took a total of only 12 days with good co-operation from the local women and the help of interpreters. It has been assumed that there are no seasonal biases in the data as the wood is used for cooking only and diets change little through the year.

RESULTS AND DISCUSSION

A sharp difference in levels of firewood consumption exists between irrigation scheme villages and riverine villages where, on average, more than twice as much firewood is consumed per person. (104.8 Kg./person/month in riverine villages; 49.6 Kg./person/month in scheme villages). Individual variations between families are very great in all villages so that the concept of an 'average' family is a little abstract but seems to be the only practicable way to discuss firewood use for predictive purposes.

Along the river, families are Moslem and in most households studied there was one, two or three wives collecting wood. This leads to very variable family size and questions had to be asked on the number of hearths being fuelled by each woman and on the number of people participating in meals served from each hearth in order to get accurate estimates of per caput firewood use. On the Hola Scheme, the dominant tribe is Pokomo but a mixture of people from different tribes live in each village. Whatever tribe a family belongs to, only one house building is allowed per family with a consequent limit on the number of hearths than can be fuelled. This limits the amount of wood that can be used, despite on average slightly larger families. On the Bura Scheme, there is also only one house per family but a greater mixing of tribes from all over Kenya. Family sizes are smaller than at Hola or than along the river for two main reasons. The first is that some tenant farmers do not want to bring their families to the scheme from their home area until the inadequate schooling for children and almost non-existent health facilities have improved. The second is that most families are young with relatively small numbers of children, as yet. This is in contrast to the scheme at Hola where families have been established for 25 years and overcrowding in many households has resulted because of lack of alternative occupations in the area. The distribution of family sizes on the two schemes and along the river is shown in Figure 10.

When levels of household firewood consumpton per day are compared (Figure 11), a marked similarity in levels of consumption can be seen on the two schemes but a much greater level of consumption is apparent along the river. In Figure 11, per caput firewood consumption per day is compared but here, higher consumption levels are apparent on the Bura Scheme compared to the Hola Scheme though both are still considerably less than consumpion levels along the river. This relates back to the number of people in each household and implies that whatever the number of people living in a household, the same amount of wood is used for cooking. Baily (1979) also comments that the consumption of wood depends more on the quantity of food cooked than on the number of people in the household. Thus, per caput wood consumption decreases with increased family size but household wood consumption changes little. This has great implications for plantation planning as this cannot be done on a 'per household' basis rather than the 'per caput' basis used in the project planning report. As the number of households on an irrigation scheme stays relatively constant even when the overall population increases, planning on a household basis makes decisions on plantation size much easier to make.

The distance travelled to collect wood and hence availability of wood is the chief factor accounting for the big difference between consumption levels on the irrigation schemes and along the river. Along the river, the women travel between 200 and 800m and the trails they use through the forest are well shaded. On the Hola Scheme, women travel from 2 to 5 km, the round trip taking up to five hours. As they collect in the dry bush, they are afforded little shade and leave early in the morning to miss the heat of the day. This is clearly a disincentive to waste wood and more frugal use is made of it. On the Bura Scheme, most women interviewed collected wood in the dry bush and not from the wood piles that has been set aside during the construction phase. They explained that during their first months at Bura they had used the piles because they were close to their villages. Subsequently they had learned, from the local women on the scheme, which species were good to cut in the bush and had stopped using the piles because much of the wood had poor burning qualities, there were too many snakes in the piles and many of the pieces of wood

58

FIGURE 10 Distribution of family sizes on the Hola Irrigation Scheme, the Bura Irrigation scheme and in riverine villages

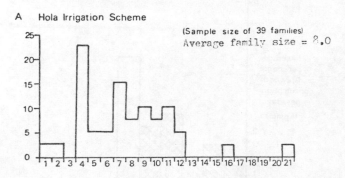

A   Hola Irrigation Scheme

(Sample size of 39 families)
Average family size = 8.0

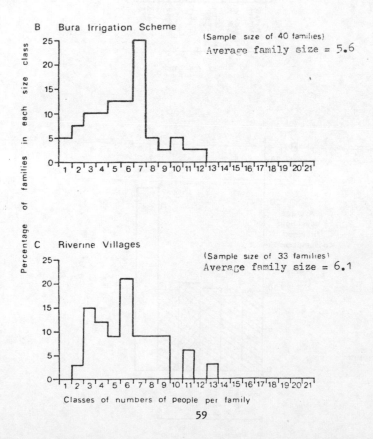

B   Bura Irrigation Scheme

(Sample size of 40 families)
Average family size = 5.6

C   Riverine Villages

(Sample size of 33 families)
Average family size = 6.1

Percentage of families in each size class

Classes of numbers of people per family

FIGURE 11 Levels of firewood consumption by household
and per caput

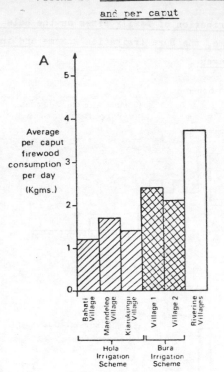

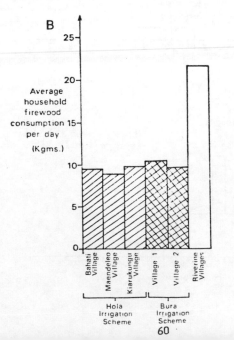

in the pile were too large for them to handle. They tended to go into the bush in small groups because most of them were unfamiliar with this type of countryside and were frightened to go out alone. Most women on the Bura Scheme walked a maximum of 2 km to collect their wood as the bush around Bura has not yet been cut extensively. It was noticeable that, on the whole, women who come from other parts of Kenya collect larger bundles of wood (around 50 Kg.) than the local women because they carry wood on their backs using headbands instead of carrying headloads (riverine and scheme village average = 29 Kg. for adults).

In summary, therefore, wood is harder to obtain, families are more nucleated and hearths are fewer on the irrigation schemes than along the river. These factors account for the differences in fuelwood consumption between the schemes and the riverine areas. It further appears than on the irrigation schemes, household firewood consumption does not significantly increase with increase in family size.

IMPLICATION OF THE RESULTS FOR PLANTATION PLANNING

The results of the firewood survey are compared with the figures used in the PPR (1977) in Table 5. The average consumption figure of 0.718 m³ per person per year (solid volume) found at Hola, is very similar to the 0.7 m³ used in the PPR (1977). At Bura, the average figure is significantly higher at 1.125 m³ per person per year. This could imply that the 4,500 ha set aside for forestry at Bura (3,906 ha of actual irrigated land and 15% to allow for roads, irrigation works and firebreaks) might be inadequate to meet the needs of the scheme. However, it has been shown above that firewood needs can be predicted using household numbers instead of per caput consumption. This means that the size of area that needs to be irrigated for forestry can be calculated by using household numbers and the productivity figures that are available from the trials at Hola.

Two reports have now been produced at Hola which give some indication of plantation productivity (Van der Veen, 1981 and Mulwa, 1982). They both state that of the 24 species originally used in trials, four are recommended for planting at Bura. These are Prosopis juliflora, Eucalyptus camaldulensis, Azadirachts indica and Cassia siamea. The charcoal yields given by these species at 20 months and 32 months growth are given in Table 6a. These figures are converted to solid volumes of wood in Table 6b in order to compare them with the figure of 15 m³ per ha per annum used in the PPR (1977).

Clearly, Prospis juliflora is outstanding in its performance and it has been recommended for use at Bura. If an average figure of 27.5 m³ per ha per annum is used and it assumed that the whole area is planted in Prosopis juliflora, then the total number of hectares that needs to be irrigated for 5,150 households is 824, managed on a rotation basis. On top of this, the administrative and commercial population, officially on the scheme and unofficially associated with it, needs to be catered for. On the whole this population will cook on charcoal and an estimated total of approximately 4,000 households fall into this category. Van der Veen (1981) estimates that each hectare of Prosopis juliflora can produce approximately 130 bags of charcoal in 20 months,

61

Table 5   Comparative Table of Wood Consumption Values

Data Source	Consumption in m^3 per person per year (solid vol.)	Consumption in kg. per person per day
1. IBRD (World Bank), 1977	0.991	2.00
2. Project Planning Report, 1977	0.700	1.41
3. Bahati (Hola Scheme), 1982	0.594	1.20
4. Maendeleo (Hola Scheme), 1982	0.842 Av. = 0.718	1.70
5. Kiarukungu (Hola Scheme), 1982	0.718	1.45
6. Village 1 (Bura Scheme), 1983	1.210	2.44
	Av. = 1.125	
7. Village 2 (Bura Scheme), 1983	1.040	2.10
8. Riverine Villages, 1982	1.852	3.74

The figures have been converted from kg. to m^3 solid volume using a specific gravity conversion factor of 0.75 given in the PPR, 1977.

Table 6a   Charcoal Yields from the Trials at Hola

Species	Charcoal Yields in Kilograms per Hectare after	
	20 months	32 months
Prosopis juliflora	3,500 (130 bags)	6,790
Eucalyptus camaldulensis	700	1,180
Azadirachta indica	650	1,070
Cassia siamea	1,000	1,490

(from Mulwa, 1982)

Table 6b   solid Volume Wood Yields from the Trials at Hola

Species	Solid volume Yields of Wood in m^3 per ha per year after	
	20 months	32 months
Prosopis juliflora	24.8	29.9
Eucalyptus camaldulensis	4.9	5.2
Azadirachta indica	4.6	4.7
Cassia siamea	7.1	6.6

A conversion ration of 84.75 kg of charcoal for 1.0 m^3 solid volume of wood was obtained from kiln specifications discussed in the PPR, 1977 and used to give Table 6b.

enough for 2.4 families per year. This means that approximately 1,700 hectares need to be irrigated for this sector of the population. A round figure of about 2 600 ha therefore needs to be irrigated for the Bura Scheme. This is less than the figure initially proposed in the PPR (1977). If, however, the plantation is only partially planted in Prosopis juliflora and equal amounts of the four recommended species are planted, then a total of about 6,000 hectares needs to be irrigated. This is rather more than is recommended in the PPR (1977).

The question of how much to plant of the four recommended species brings to the fore the question of species preference. In this matter, the importance of local involvement, especially that of the women cannot be overstressed. The survey in the irrigation scheme villages and along the river indicated a marked species preference as long as there was a choice. Three species were commonly chosen by women along the river, these were Acacia elatior, Cordia gharaf and Lecaniodiscus fraxinfolius with occasional Terminalia species. The first two are dense woods, suitable for slow burning and the third is very fast burning and suitable for kindling. On the irrigation schemes a different three secies were commonly collected. These were Acacia reficiens, Terminalia orbicularis and Cordia ovalis which are all species found in the dry bush. Apparently small considerations like the number of thorns of a bush readily turn people away from collecting it. Prosopis juliflora has a dense covering of thorns on its multiple trunks and the foresters at Hola comment on the difficulties of harvesting it and the dislike of this job by those who have to do it. Brokensha and Riley (1978) also comment on species preference for different uses. Consultation about these matters with the farming families could avert some costly mistakes. Women's involvement in the plantation upkeep might be an efficient way of ensuring plantation productivity if they can see the benefits of having a guaranteed and accessible source of firewood.

CONCLUSIONS

Despite concern for fuel supplies at the start of the Bura Scheme, the planned firewood plantation has not yet been implemented, even though some tenant farmers have been living on the scheme for over 18 months. Reasons for this delay have been listed above. This study fills in some of the information gaps for plantation planning. Recent reports on the irrigated plantations at Bura have come from the Government of Finland (Hakkila et al, 1982) which has drawn up an implementation plan for the plantation and is expected to fund approximately 11.5% of the total cost and may fund considerably more. If all goes to plan, the first harvest of firewood can now be expected around the end of 1986. In the interim, wholesale destruction of bush and riverine forest can be expected with considerable hardship for the women who will have to go further and further afield to collect their wood.

The provision of firewood is a vital element in irrigation scheme planning. Without firewood plantations farming families must depend on local sources of natural vegetation which can be rapidly exhausted and which may (as in the case of the Tana River floodplain forest) already be used by resident peoples and be of conservation importance.

Proper and timely surveys of fuelwood resources and needs, and properly established and maintained pilot schemes are needed. Where appropriate, aid agencies should assume responsibility for seeing that such work is done.

REFERENCES

Bhatt, Chadi Prasad (1981) Trees - A source of energy for village dwellers. Paper presented to the NGO Forum at the UN Conference on New and Renewable Sources of Energy, Nairobi 1981.

Baily, J. (1979) Firewood use in a Sri Lankan village: A preliminary survey. Occ. Pap. on Appropriate Technology, School of eng. Sci., Univ. of Edinburgh.

Brokensha, D. and Riley, B., (1978) Forest, foraging, fences and fuel in a marginal area of Kenya. Prepared for USAID Africa Bureau, Firewood Workshop, Washington.

Eckholm, E. (1975) The other energy crisis: Firewood. Worldwatch Paper 1 Worldwatch institute, September 1975.

Hoskins, M.W. (1980) Community forestry depends on women. Unasylva 32 (130): 27-32.

IBRD (1977) Bura Irrigation Settlement Project Appriasal Report (Yellow Cover) Washington.

Kamweti, D.M. (1981) An overview: Fuelwood and charcoal/tree planting for energy in Kenya. Paper presented to the NGO Forum at the UN conference on New and Renewable Sources of Energy, Nairobi, 1981.

Mulwa, R.S. (1982) Summary of the results of experimental work carried out in Umoja Forestry trials from December 1979 to April 1982. Vol. I. Unpublished report from the Hola Irrigation Research Station, Kenya.

Openshaw, K. (1981) Woodfuel and the energy crisis: Problems and possible solutions. Paper presented to the Panel on fuelwood and charcoal at the UN conference on New and Renewable Sources of energy, Nairobi 1981.

Project Planning Report - Bura Irrigation Settlement Project (1977) by Sir M MacDonald & Partners, Hunting Technical Services Ltd. and East African Engineering Consultants for the National Irrigation Board, Ministry of Agriculture, Government of Kenya.

Shakow, D., Weiner, D. and O'Keefe, P. (1981) Energy and Development: The case of Kenya. Ambio 10: 206-210.UN (1981).

Report of the Technical Panel on Fuelwood and Charcoal to the UN conference on New and Renewable Sources of Energy, Nairobi, 1981.

Van der Veen, E. (1981) Forest Area, Bura. Unpublished report from the Hola irrigation research station, National Irrigation Board of Kenya.

66

HAUSA WOMEN AND AGRICULTURAL CHANGE
ON THE KANO RIVER PROJECT, NIGERIA

C. Jackson

The Kano River Project, (KRP) is a large scale irrigation scheme situated some 50 km south of Kano city. The area is flat, seasonally dry with high rural population densities (178 per sq. km) and characterized by relatively small and intensively cultivated farms on which only simple hand tools are used. The population is overwhelmingly muslim but pockets of pagan Hausa remain and were included in this study which was conducted between 1976-78 in both on and off project villages. A more detailed account of the study can be found in Jackson, (1981). The KRP, which began in 1970, is basically a canal irrigation scheme which produces mainly wheat and tomatoes. Articulated project objectives include increased food production, the creation of additional employment opportunities and the raising of the standard of living. No provision was made for the participation of women in this scheme.

The study focussed on the different effects the scheme has had on muslim women and pagan women, and something should be said on this before we look specifically at agricultural change. Muslim women are largely concerned with making snack foods for sale and petty trading whilst pagan women are own-account farmers. Yet the effects of the KRP on pagan women have been profoundly negative whilst muslim women on the KRP have largely prospered through their non-agricultural occupations. Pagan women compete with men for the same set of productive resources - land and labour - and are less and less able to make farming financially worthwhile in a situation where land is costly, labour scarce, fertilizers hard to gain access to and where their labour time is increasingly absorbed in unpaid household farm work. Muslim women, on the other hand do not depend on husbands to allocate productive resources, for in their petty commodity production they use only every day domestic utensils, depend on no-one else's labour, their income derived is entirely personal and consequently they have little committment to the household as a unit of production. Their independence is increasing on the KRP with greater financial freedom and in spite of, perhaps because of, increasing seclusion.

PAGAN HAUSA WOMEN

It is interesting to observe that although pagan women are all own-account farmers, fewer pagan women own land than muslim women. In a survey of 147 muslim men and 24 pagan men of the KRP area, I found that some 25% of muslim men reported farms owned by their wives but only 4% of pagan men had wives that owned land. Similarly, it was found that of all the farms worked by the 47 pagan women of Yan Tomo village only 8% were permanently owned i.e. either bought, inherited or gift farms. Pagan women do not as a rule inherit land and generally pagan women are not land owners. So, curiously, pagan women were less directly affected by the land redistribution on the KRP than

67

muslim women were.

The most common form of tenure for pagan women is the farm loaned from her husband (67% of farms) followed by farms loaned by her natal family (11%) and from other sources (12%). A pagan husband is obliged to allocate a small plot to his wife for her personal use. So pagan women are clearly dependent on marriage for their access to the means of production - a dependence which is becoming greater as a consequence of the KRP because rapidly rising land values have made it even more difficult for pagan women to buy or borrow farms on the 'open market'.

The average pagan woman's holding on the KRP is about 0.6 ha and consists of 2.2 plots of land. Farm sizes generally have been on the decline for many years in this part of Hausaland, because of the population densities, and the KRP has exacerbated this trend through the outright confiscation of a certain amount of land required for the large 'government farm' etc., and the reduction in size by some 10% of all reallocated holdings to provide land for canals and roads.

The yields of women's farms are markedly lower than those of farms worked by the household.

These are wet season crops. In the dry season women lose the use of their plots if they are irrigable, for only 2 Yan Tomo women attempted irrigated agriculture during the period of my study. Some 60% of women's farms were irrigated and it can be seen that Yan Tomo pagan men have broken the custom of providing a wife with a farm for her use alone: a wife's rights now only hold in the wet season. This is not the case with male private plots (e.g. those allocated to junior men within the household) and clearly women's plots are seen as the least important and the most dispensable from the household head's viewpoint.

Pagan women have been excluded from the opportunites of irrigated agriculture as own account farmers, their sole relationship to it being as household workers on the farms of husbands and sons.

The reasons for the low yields on women's farms stem largely from the attitude of men to these plots; when allocating land the household head is likely to give women the least desirable land, i.e. land with poor soil or water logging problems, distant farms and so on. It was found that only 1 of 47 women managed to get Fulani cattle corralled on her farm, only 17% obtained any manure or compound sweepings, only 6% used chemical fertilizers, and none belong to the Co-operative. Pagan women lack the influence and the money necessary to obtain adequate farm inputs and consequently yields are low.

A similar picture of tiny farms and inadequate manure is given by P. Roberts (1979), writing about Hausa women in Niger and these problems are not entirely the consequence of the KRP.

However the KRP has directly generated another obstacle for pagan women - the wage of 2.00 Naira per day for hired labour (set by the KRP) is beyond their reach and has made it difficult for women to obtain sufficient labour inputs for their farms.(£1 = N1.2 in Sept. 1981) Furthermore, women complain that they are now required to work much longer on their husbands' farms and therefore have less spare time for their own farms. Pagan women are obliged to work on the household farms during the mornings of Saturdays to Thursday and

Fridays and afternoons are their "spare time" for cultivating their own farms. But they say that before the KRP a morning's work ended at 12 noon, now it ends at 2 pm or later. Women also complain that because of the widespread destruction of trees on the KRP it takes them very much longer to gather firewood than it used to and this too reduces their "spare time".

Not only do women have less of their own labour times for their farms, but they are increasingly unable to get unpaid or semi-paid labour either; neighbours, friends and relatives are also pressed for time and children are in school. Given this situation where women have less and less of their own time, lack resources to hire labour or the ability to call on traditional 'help' relationships for farm work, we can expect to see yields dropping even further and certainly not the remotest possibility of women increasing their productivity, adopting new crops and techniques, venturing into irrigated agriculture, raising their incomes or any other of the goals of the KRP.

Apart from their own farms pagan women are very much part of the farming unit and responsible for much of the farm work of all sorts. A wife is a more likely farming partner for a pagan man than a brother; brothers were mentioned in 31% of muslim replies and 12% of pagan replies to this set of questions. However women featured in 44% of muslim and 61% of pagan replies to questions about farm work.

It was found that pagan women are less commonly paid for threshing than muslim women are, and amongst pagan women, wives are the least rewarded of all. Men who were not married had to give 1 or 2 bowls of grain plus 20 to 30 kobo to induce other compound women to thresh their grain, whilst all the wives doing this work were unpaid.

MUSLIM HAUSA WOMEN

It is commonly assumed that muslim Hausa women do virtually no farm work - for example, David Norman's influential studies (1972) calculated that less than 1% of farm labour was contributed by women. Whilst very few muslim women are own-account farmers, they do engage in agricultural work on their husbands' and sons' farms and participate as wage labourers.

A study of reports from Sokoto, Kano and Zaria provinces from 1900 to 1930 revealed however that this was not the case in the fairly recent past. Many examples were found showing that women farmed on their own account and worked as full gandu members. (Gandu is, amongst other things, the ideal family farming unit. Under the leadership of the senior male its members work the household farms during the mornings and are allocated private plots for personal use which are cultivated in the afternoons and on Fridays). For descriptions of gandu see Hill (1972) and Wallace (1979). This was substantiated in interviews with a number of old women in Yan Tomo who independently gave accounts of working in a similar way to that of pagan women today. These women remember the past with mixed feelings. On the one hand, women used to be given a small share of the crops from the gandu farms on which they worked, and women also used to have their own granaries and considerable access to grain. On the other hand, it is agreed that women also used to work much harder, they were paid much less for their harvesting work, and they were obliged to contribute to the household food supplies to a much greater extent than they do now.

There seems to have been a long term pattern of change whereby women were transformed from perpetual juniors-in-gandu to independent petty commodity producers engaged largely in extra household economic activities and participatng in agriculture mainly as paid labourers.

Without ascribing causality, one can however see that increasing pressure on land is likely to have reduced the size of women's plots to insignificance, and made labour relatively plentiful and thus women's labour dispensable to some extent, whilst at the same time seclusion was spreading and the deepening monetarization of the economy opening up other opportunities for women.

The question of the specific effects of the KRP on the quantity of farm work done by women is difficult because of the lack of adequate information on women's farm work in the pre-KRP situation. Most men aspire to an ideal to extreme wife seclusion where no farm work is done by their wives and thus surveys of men usually severely under-estimated female farm work. And of course the situation will be different for different categories of women. But there are important qualitative aspects of changing farm work patterns as well to complicate the issue; unpaid family farm work must be distinguished from paid family farm work and wage labouring.

Certainly, there is more farm work available now which some women have taken advantage of, the older unsecluded women and non-secluded farmstead dwelling women. But other categories of women do less family farmwork now than ever before; this change has been especially clear cut for women who previously lived in dispersed farmsteads but were resettled by the KRP into nucleated villages such as Yan Tomo. Even within the time scale of this study, women who had done certain farm tasks in 1976 were seen to withdraw from them in 1977.

How important are women to the production of wheat? A careful frequent visit survey of 10 Yan Tomo farmers over two seasons showed the extent of women's work in wheat harvesting. (Data collected by Dr. Richard Palmer-Jones) Unfortunately the hired labour category was not broken down by sex but a considerable proportion would have been women. It can be seen that hired labour is required on a large scale to deal with the wheat harvest, only about 30% of labour being family labour. However, in this 30% women are somewhat more important than men. Add to this the preponderance of women in the hired labour category (especially as winnowers) and we see that women are playing an important part in wheat production. We must point out though that it is misleading to group compound women with family labour as few of the former are unrewarded for their work. The involvement of women in other farm activities is shown in Table 7, which indictes their importance in harvesting, and Table 8 which shows that they participate in some of the lowest paid tasks. Threshing rewards are higher than winnowing becaue both sexes take part, but winnowing is paid at less than half the rate of other jobs. We can see why women winnowers preferred to be paid in kind. Threshing on the farm is often done by men.

Table 7   Pagan Women's Farm Yields

| | Estimated Yield | |
	Women's Farms	Household Farms
Sorghum	503	651 Kg/ha
Millet	257	616 Kg/ha
Groundnuts	0.43	2.1 bags/ha/head
Beans	0.31	0.63 bags/ha/head

Table 8   Wheat Harvest Labour, Yan Tomo

		% of mandays 1975-76 & 1976-77*
Threshing	Self	9%
	Compound men	6%
	" women	2%
	Hired Labour	73%
Winnowing	Self	2%
	Compound men	3%
	" women	25%
	Hired Labour	70%

Where threshing and winnowing were recorded together the mandays have been halved.

*Because there are few records for 1976-77 they have been amalgamated here.   71

DRUDGERY AND WOMEN

By drudgery we mean unrewarded domestic work and before closing we should compare the relative situations of pagan and muslim women on the KRP.

As we have pointed out, muslim women participate in little unpaid farm work, unlike their pagan sisters. Similarly there seems to be a process of commoditization of domestic labour occurring in muslim households. In the past women were involved more in wood and water collection, in house building (floor making in particular) and in crop transport and the collection of "bush foods". Now however, men are usually responsible for wood and water collection or purchase these, cement floors do not require women's work, bicycles are becoming common as means of crop transport and less collected foods are consumed. Whilst these previously female tasks are being displaced I would argue that rather than depriving women of status earning activities this is indeed merely relieving the drudgery of muslim women's lives, for status is clearly not based on the contribution to unrewarded household labour in Hausaland.

The KRP surveys found that muslim men are responsible for over half of all laundry tasks and most of the weed collection. Threshing within the compound however, is exclusively women's work. It is carried out all year round but there is a peak in threshing in the post-harvest period. Most (62%) of the Yan Tomo women were not paid for threshing for their husbands but a sizeable minority (38%) were. This seems by all accounts to be a new practice which originated with the wheat threshing, for which women demanded payment on the grounds that male wage labour was paid for this work and that the crop was not grown for household consumption. By analogy, surplus millet, sorghum and maize being threshed for sale by compound women is also often paid for.

Planting is a task which is clearly unrewarded and in which women are involved less and less. Only 11% of my sample reported doing any planting and many made unsolicited remarks to the effect that before they used to plant but in new Yan Tomo they do not.

Grinding of grain is now usually done by the mills in the villages thereby saving countless hours of unpaid women's work.

There are one or two areas where KRP muslim women seem to be experiencing a heavier load of unpaid domestic work, e.g. in fodder collection, because goats and sheep must be kept in the compound for two seasons, although there are plentiful grasses alongside canals. There also seems to be more cooking to be done when households are engaged in irrigated agriculture because it is the custom to cook twice a day during an agricultural season and only once a day otherwise.

However it is the pagan KRP women who are not only involved in much greater farm drudgery but also now find themselves with additional fodder collection chores as well as the problem of finding firewood to cook with when most of the trees have been destroyed in the building of the KRP.

Their expeditions to collect wood frequently take all day and involve very long walks and heavy loads. Pagan women are not paid for their off-farm processing activities to the extent of muslims and the threshing, winnowing and bagging of wheat is an onerous task which off-project pagan women are free of.

Domestic labour has been described by Young, (1979) as "a particularly rigid feature of the sexual division of labour". Yet it has shown considerable changes in Hausaland and on the KRP and should not be taken as a "given" which cannot be changed by planners.

To summarize a rather complex picture: the KRP has had dire consequences for the farming of pagan Hausa women. Their access to land has been reduced because of the inflated value of irrigated land, the yields on their plots have suffered because of their increasing work obligations on gandu farms and their inability to compete for increasingly expensive farm inputs, they have been excluded from the opportunity of farming irrigated land, and have become involved in a greater amount of unpaid labour, i.e. drudgery. On the other hand, most muslim Hausa women are less involved in unpaid agricultural work than before the KRP, although they engage in a greater quantity of rewarded farm work through the double season cropping and as a result of the trend towards payment for crop processing.

The independence of pagan women farmers is more apparent than real and their dependence on husbands for land and relatives and friends for farm labour and the ability to compete with men for purchased farm inputs has made them vulnerable to the negative effects of large-scale irrigation. Muslim women have but a tenuous connection with the household as a unit of production, the idea of seclusion frees them from the increasing demands of unpaid labour, they have benefitted from the greater availability of paid farm work, and the extreme separation of the muslim male and female economies means that women are not competing with men for the same set of increasingly scarce productive resources.

REFERENCES

Hill P. (1972) Rural Hausa: a village and a setting. Cambridge University Press.

Jackson C. (1981) "Change and Rural Hausa Women: A study in Kura and Rano Districts, Northern Nigeria" University of London Thesis submitted for Ph.D.)

Roberts P. (1979) "The integration of women into the Development Process - Some conceptual problems" Bulletin of the Institute of Development studies, report on conference 133.

Wallace C. (1979) Rural Development through irrigation: studies in a town on the Kano River Project. Centre for social and Economic research, Report No. 3, A.B.U.)

Young, K. (1979) (ed.) The continuing subordination of women in the Development Process Bulletin of the Institute of Development Studies. Conference No. 133 p. 14.

ACKNOWLEDGEMENTS

I would like to acknowledge the assistance given to the study by the Overseas Development Administration, Dr. George Abalu and the Dept. of Agricultural Economics and Rural Sociology of A.B.U., Dr. Tina Wallace, Dr. Richard Palmer-Jones, Dr. Ian Carruthers and Jeremy Jackson.

THE LAND ISSUE IN LARGE SCALE IRRIGATION PROJECTS:
SOME PROBLEMS FROM NORTHERN NIGERIA

A.C. Bird

INTRODUCTION

The aim of this paper is to illustrate the broad nature of the problems associated with land in the construction of large scale, formal irrigation projects in Northern Nigeria. It contains general points but is based upon practical experience in designing and supervising surveys for land acquisition purposes. Some specific points will be raised concerning technical problems encountered on these two projects and finally a brief comment will be made so as to place the "lands" issue into perspective with other problems being encountered in the construction and management of such schemes in Northern Nigeria.

BACKGROUND TO THE PROJECTS

Development of large scale, formal irrigation schemes in Northern Nigeria is generally handled by the Federal Government Ministry of Water resources through the River Basin Development Authorities. These ten Authorities were set up by military decree in 1976 (although two did exist previously) with the aim of developing Nigeria's water resources for the increased benefit of the nation by utilizing the revenues from oil exports to build schemes both for hydro power generation and irrigation. They now come under the Federal Government's "Green Revolution" programme which aims to dramatically increase food production in an attempt to cut rapidly increasing food imports. Construction of the Bakolori project is near completion and work on the Dam at Dadin Kowa is well advanced with first impounding due in the rains of 1983. The irrigation construction contract at Dadin Kowa is about to be let.

The main elements of these projects are a dam and reservoir, permanent works in the irrigation area and the actual irrigated fields (generally expressed as the net irrigated area).

Some Nigerian projects do not have all these elements (e.g. the South Chad Irrigation project utilizes water from Lake Chad fed by a canal, and the Goronyo dam project on the Rima in Sokoto State is purely a storage dam for irrigation a long way downstream), but it should be kept in mind that each of these elements requires different treatment from the land acquisition point of view.

The dam site and reservoir are permanent works and all the land beneath them is of course lost for cultivation purposes in the future. By law (the 1978 Land Use Act) compensation has to be given for these losses. The irrigation areas are more complex as only part of the land (up to 25% in some cases) is taken up by constructed permanent works (e.g. roads, canals, drains, etc), while the remainder is developed as irrigated land capable of being cultivated all year round.

LAND POLICY AND CONSTRAINTS UPON IT

Policy with regard to land is constrained by present land law and by decisions already taken (be they taken consciously or not) about the nature of the development of the project at feasibility and design stage. At Bakolori there was no conscious formulation of land policy; it just evolved by default as each crisis arose, with the result that the short term "easiest" expedient routes were taken. At Dadin Kowa the feasibility study and the designs omitted any mention of land management in the irrigation project area. However, as the acquisition surveys have been carried out before the construction contract has been let there is still just time to formulate a land policy for the irrigation area.

Ideally, land policy needs to be fixed before surveys for land acquisition take place, as the design of the survey requires to be specific to the chosen policy. Factors constraining land policy include land law, cost, engineering criteria already laid down, ease of operation of the project and socio-economic factors. In the past, over-emphasis has been placed on engineering criteria, to the detriment of other factors, particularly socio-economic factors. These have been very badly undervalued, especially when considering their real importance. The law and political factors are of course very relevant and perhaps form the overall basis of land control.

LAND LAW BACKGROUND

The Land Use Decree (now Act) of 1978 vests control over land to the relevant State Governors, who, with local governments in rural areas, are empowered to recognise and grant Rights of Occupancy. These Rights of Occupancy can either be customary or statutory. Customary Rights of Occupancy confer exclusive possession for an indeterminate term and are inheritable; however, it is generally recognised that dealings in land of a variety of kinds do take place and have done for a long time, although there may seldom be any written record of them. In these circumstances, such rights have acquired a marketable value to the extent that the right holder's estate amounts to that of freehold. Statutory rights of Occupancy are registered with the appropriate authority (state or local governments, depending upon whether it is urban or rural land) and cannot be inherited or bought or sold. They are held on leases (99 years in many cases) for which a minimal ground rent is paid, but actual registration costs in the order of 50 Naira.

Land cannot be expropriated, except according to the provisions of the Land Use Act. Only State Governors (and in rural areas local Governments) are authorised under the Act to revoke (i.e. compulsorily extinguish) private rights in land. Federal Government agencies cannot do this without authorisation from the State Governor, and following such authorisation must adopt the procedures laid down in the Act with regard to compensation. When such Rights of Occupancy have been revoked, either alternative equivalent land has to be given to the former right holders, or compensation becomes payable under Section 29 of the Act. In principle compensation is for "unexhausted improvements", that is "any thing attached to the land clearly resulting from expenditure of capital on land by an occupier . . . and improving the productive capacity". Each State Government has issued its own circulars with regard to this and there are different interpretations of the Act leading to inconsistancies between one State and another. In the case of the Dadin Kowa project, which straddled three State boundaries, this was potentially a serious problem.

COMPENSATION

Some State Governments (e.g. Bauchi) actually refer to compensation for "land" as opposed to improvements to it, and have different cash rates for compensation according to land type and location. Others, e.g. Borno State, do not have any rates and one can only assume that, following the Land Use Act, equivalent replacement land is given. In effect then, cash compensation is actually paid in order to obtain the Rights of Occupancy to land as well as for any improvements which appear on it. These improvements incude economic trees, shadufs, field wells, farm fences and any standing crops and in built-up areas any buildings and amenities, both public and private. In areas where land is only temporarily acquired and then handed back to occupiers at a later date (e.g. in irrigation areas) disturbance compensation should be paid (according to the 1962 Land Tenure Law), plus compensation for any improvements destroyed.

Rates of compensation vary greatly from one State to another; however, they can be considerable and the cost of compensation, especially for acquiring the Rights of Occupancy, is in itself a great constraint upon land policy selection. Unfortunately, the cost of compensation is very rarely considered at the feasibility stage of any of the projects, possibly as compensation costs, despite being high, are still a small proportion of the enormous cost of constructing formal irrigation schemes in Northern Nigeria. As a guideline, developing land for irrigation costs about 10,000 Naira per hectare, not including the cost of any dam or reservoir construction, or compensation costs for either the reservoir area or the irrigation area.

Compensation costs in rural riverine areas can average around 1,500 Naira per hectare for acquisition of the Rights of Occupancy and for improvements to land. The cost can be considerably higher if households and amenities have to be compensated, e.g. as in reservoir areas. The biggest problem, however, is that they are not included as part of the construction costs of projects, and there is often very serious difficulty in making such money available, especially when compared with the ease with which foreign loans for the construction of projects have recently been obtained. When funds do arrive for compensation payment it is often too late to distribute them in a sensible phased payment programme that would lessen the risk of them being frittered away.

PUBLIC PARTICIPATION

Another recent constraint on land policy is that, particularly after the emergence of civilian rule in Nigeria from October 1979, projects were to a far greater extent open to political pressure, especially at State and local level. They must be made acceptable to the local population, and some degree of public participation should be envisaged, even if it is only informing people of what is going on. This was learnt at great cost to human life at Bakolori, as well as considerable financial cost to the Government in contractors' claims, when the farmers blockaded the site and stopped the contractor working for a considerable time. This episode culminated in armed riot police being used to "retake" the project area with the loss of at least 50 and possibly around 200 lives. (See Bird in press)

SOCIO-ECONOMIC FACTORS

Socio-economic considerations should take a high priority in project planning, but unfortunately, these extremely fundamental aspects have been very seriously undervalued. Over-emphasis has been placed on the purely technical engineering aspects of the design and construction to the serious detriment of the future of the projects. Even basic agricultural and soil survey work has been undervalued and feasibility studies have been carried out too late, often to justify decisions already taken (in the case of Bakolori, the formal presentation of the Impresit report occurred after the turn-key design and construction contract had been awarded). Furthermore, their scope was too narrow, concentrating on the immediate project area and not putting the project into the context of the whole river basin system. Externalities and detrimental aspects were glossed over and emphasis laid on the positive aspects, rather than pointing out problems and suggesting ways of avoiding, overcoming, or minimising them.

With regard to land, the basic "land equation" does not even balance at either Bakolori or Dadin Kowa. At Bakolori the land area lost under the reservoir is approximately 7,200 ha, the irrigation area covers 30,000 ha gross, but at least 25% of this is lost under permanent works. In addition, studies indicate that up to 20,000 ha of riverside land downstream of the project is now seriously losing its productive capacity, as waters previously used are now diverted into the irrigation area and the dam attenuates the peak floods which before kept the riverine lowlands wet throughout the dry season. (Adams 1981, 1983) Put simplistically, 34,700 ha have been lost to cultivation in order to irrigate 22,500 ha. The increased production required from irrigation to compensate for this land loss is mind-boggling, irrespective of the project construction costs (350 million Naira) and the resettlement cost for the reservoir. The situation at Dadin Kowa is even less favourable with a reservoir of 35,000 ha displacing 23,000 people suppplying an irrigation area of 25,000 ha. Some of the irrigation area is densely populated, as 12,000 people have already been resettled into it, following the flooding of their land and homes by the Kiri dam further downstream.

Things are made worse by the fact that very few projects have had representative pilot farms set up before-hand and very few have had adequate baseline socio-economic surveys carried out. The Bakolori final report (Impresit 1974) used a sample of 41 unrepresentative farmers for an area of 30,000 ha. The report also badly misjudged the percentage of the project area occupied, assuming 60% occupancy, when even a brief glance at the 1962 air photographs indicates the figure is nearer 95%.

As stated previously, the engineering aspects have been given inordinately great precedence over land policy. The fundamental issue here is that formal large scale irrigation schemes, regardless of whether they are sprinkler or surface distribution, are generally laid out on a rectilinear pattern. In northern Nigeria the riverine areas are by their very nature heavily populated and the land densely occupied. The existing land holding pattern is highly irregular in shape and pattern and has developed over a long period, influenced by long term social, economic and environmental factors. At first sight this may seem, to an unfamiliar eye, highly inefficient; however, closer study reveals that the situation is in a very complex, finely balanced equilibrium and is extremely rational and efficient at household level in social and economic terms.

The introduction of formal irrigated agriculture results in this pattern having to be destroyed and a new rectilinear pattern imposed to fit in with the irrigated field layout. The new fields can only be practically managed in unit sizes, according to water distribution design, generally a strip. The introduction of formal irrigation therefore results in the old land tenure pattern being irrevocably destroyed. A new pattern has to be established which follows the selected land policy, be it re-allocated land on an equivalent area or quality basis, or direct control by the implementing authority.

TNE NEED FOR CONTROL OVER LAND FOR IRRIGATION PURPOSES

Different land policy options will of course allow the executing authority to have different degrees of control and regulation over the land in the project area. Experiences both at Bakolori and the Kano River project (see Wallace 1979), which has been in operation about two years longer than Bakolori, indicate that where a policy of not acquiring the Rights of Occupancy but rather re-allocating the land back to the farmers, has been followed, then the resulting lack of control over the land is a very serious impediment to operating the schemes at their optimum level. Another problem which requires control over land, is the selection of crop type, especially in small-holder aras with tiny plot sizes (average plot sizes at Bakolori prior to the construction of the scheme were 0.23 ha), as watering requirements for different crops are fixed and varied. Planting dates also have to be synchronised, which again in small-holder areas is very difficult to manage. Regulation is also likely to be desired over land transaction, especially plot subdivision, amalgamation, subletting/leasing and the buying and selling of land. This regulation is impossible to enforce, unless the Rights of Occupancy have been formerly acquired.

There appears to be a tendency for richer urban-based people or the larger farmers to buy up the land of the smaller ones who are worst hit by the disruption during the construction of the project, and who are also least able to afford the expensive inputs necessary to maximise returns fom dry season irrigation. There appears to be a move, especially near urban areas, towards farms being held by absentee landlords and being worked by wage paid labour. This illustrates the need for some degree of control over land by the implementing authority in order effectively to manage the scheme to its, inappropriate as it may be, designed optimum.

POLICY OPTIONS: ACQUIRING RIGHTS OF OCCUPANCY

Policy options for land management in irrigation areas can be broadly split between those which entail the acquisition of the rights of Occupancy over the project area by paying cash compensation or providing alternative replacement land, and those where land is temporarily acquired for construction purposes (disturbance compensation having been paid for each season the occupier is prevented from cropping), with the land being re-allocated back to the farmers at a later date.

The advantages of acquisition of the Rights of Occupancy are that the Authority has control over the land and can hence determine crop types, land preparation and planting times and watering schedules. In short, it has overall control of the operation of the scheme. The disadvantages are the cost of paying compensation for acquiring the Rights of Occupancy, which can be up to 2,000 Naira per hectare in some areas. Whilst this may seem expensive, when compared with project construction costs of around 10,000 Naira per ha it is relatively cheap. Another significant disadvantage is the political difficulty of acquiring land from peasant farmers who have occupied them for a significant time, especially if they are losing all their farmlands.

Payment of cash compensation is not likely to alleviate the long-term problems of deprived livelihoods, and in the short-term is likely to create local price inflation. The result is that a community which was originally self-sufficient in food stuffs and often selling a surplus, is not able to grow its own food. The alternative policy, as indicated in the Land Use Act, of giving replacement equivalent land outside the project area is not likely to be a viable option, owing to the land occupancy pressures in areas technically suitable for formal irrigation schemes.

REALLOCATING PROJECT LANDS

The major difficulties with the alternative policy of allocating the project lands back to the farmers after the construction of the project, thus avoiding the need to pay out cash compensation for acquiring the Rights of Occupancy, centre around the problems of deciding the criteria for defining "equivalent" land. A further problem occurs in densely occupied areas where the amount of land lost under permanent works (anything up to 20% in surface irrigation areas) exceeds the amount of unclaimed land in the area prior to the construction of the project. It may be necessary in such cases to reduce the size of all reallocated plots to overcome this.

It is also difficult to administer a programme of temporary land acquisition of this type, as it requires an efficiently run and effective Lands Office (see MRT 1980) with skilled personnel and back-up. Any delay in allowing a contractor access to land for construction purposes brought about by difficulties in land administration, could make the Authority liable for claims by the Contractor. These could be, and at Bakolori were, considerable. These are in addition to the difficulties created by lack of control over land and resulting problems of operating the scheme after construction, including the inability of the Authority to control land speculation or force farmers to carry out irrigation, grow specific crops at certain times, or carry out mechanised operations in a practical manner.

MANAGING THE LAND

Within these two major policy options there are sub-options, depending upon who operates the scheme in the case for formal acquisition policy, or how the land re-allocation is carried out if the rights of Occupancy are not acquired. If the Rights of Occupancy are acquired for the project lands, then the Authority can either farm the land directly itself, get a commercial operator to do this for him, or lease it to small-holders in unit blocks. The advantages of the Authority farming the project directly are that it can have complete control of the scheme, determining crop types, planting times and watering schedules. In theory, this should result in the most efficient manner of operation for formal irrigation schemes, allowing extensive mechanisation. However, in practice it would involve the Authority in heavy commitments of scarce resources and the record of such State-run schemes in less-developed countries is not good, partly due to the fact that wage paid labour, unlike peasant farmers, has little vested interest in the outcome of their work with resulting low productivity. In national terms it also makes greatest demands on the skills and technology which are most lacking.

If the project was worked as a commercial operation then some of the problems of low productivity could be overcome; however, the prime reasoning behind such an operation would be to make a financial profit for the operators and this is likely to be at odds with stated government objectives for agriculture and rural development. It could also be difficult to politically justify the handing-over of a government project to a comercial enterprise.

Leasing the land back to small-holders has the great advantage of utilizing existing local agricultural skills and placing less demands on the Authority's scarce resources. It also greatly lessens the problems of depriving existing farmers of their livelihood. The Authority would still have control of the land by leasing it out and have the power to terminate leases for poor or non-farming occupants. It could also form the foundation for the establishment of local co-operatives or farmers associations. However, it could be difficult to operate such a policy and great care would be required in the drawing-up and operating of the lease system. Revenue collection for water rates and inputs could also be carried out in quite an efficient manner. However, the biggest obstacle to this policy wuld seem to be its legality following the RBDA's decree No. 81 of 1979, which omits the section that previously allowed RBDA's to do this.

If the Authority chose not to acquire the rights of Occupancy and favoured land re-allocation, then either formal or informal land re-allocation could be carried out. The difficulty with formal re-allocation is that it is very time-consuming, expensive, complex and prone to error. (For details of the mechanics of this operation see MRT 1981.) At Bakolori, people were prevented from reoccupying prepared land because of the delays in implementing it, specifically pegging out and being shown their new plots. Informal re-allocation of land by local government village heads is quick, but without doubt contrary to the Land Use Act and is also open to serious abuse by unscrupulous elements. The evolved policy in the irrigation area at Bakolori was to formally re-allocate the irrigated farm plots back to the farmers after construction, using the argument that a re-allocated plot, although smaller in size (due to the land losses under permanent works and also the need to re-allocate a minimum plot size which was bigger than some of the existing small plots) was at least equivalent to the old plot, because it was improved by irrigation and could be cropped all year round. Formal re-allocaton was found to be so complex and time consuming that it has now been temporarily abandoned in some places and a policy of informal re-allocation adopted. Re-allocating land avoided the need to pay cash compensation for the Rights of Occupancy, which would have been expensive; however, the result is that management of the scheme is now extremely difficult, and the Authority can only rely upon demonstrating to the farmers that irrigated farming is an economic proposition and hoping that by example they will adopt it. This argument assumes that it is an economic proposition for small holders, a fact which has not yet been convincingly demonstrated and proven.

LAND SURVEYS

The techniques of survey for land acquisition and the methodology used at Bakolori are described in Bird (1981). Briefly, the work entailed plotting, in the field, the farm plot boundaries on copies of 1:2,500 air photographs or controlled mosaics. Improvements to the plot were then registered, and in the case of Dadin Kowa reservoir survey, a household and amenity enumeration was also carried out. The areas of farm plots were then measured or calculated in the drawing office at a later date.

Problems have been experienced with mosaic quality, both in terms of lack of sufficient ground control (this is especially a problem in unmapped reservoir sites), poor resolution and suitability due to age and time of year flown (early dry season is best). The great volume of data collected creates difficulties of data handling storage, although a manual system is without doubt the most appropriate.

However, the biggest constraint has been the need to have a firm indication of land policy before designing the survey, to avoid the collection of needless data or the omission of data required for the chosen policy. Thus it is important that the items to be compensated for must be agreed upon beforehand. There were problems at Bakolori due to the lack of a defined land policy, when attempts were made half way through the project to consolidate scattered farm plots held by the same person (see MRT 1979) into single irrigated holdings. This proved impossible, owing to the unsuitability of the survey data for this purpose.

ADDITIONAL FACTORS

Serious problems have occured at Bakolori and seem likely to occur at Dadin Kowa, due to the designs being carried out remote from the site (in Italy) and with very little detailed site knowledge. The designs are inappropriate to the areas and insensitive to local conditions. It could be possible to design a more appropriate informal irrigation system (eg. utilizing low pressure pipe delivery) which feeds water to the existing farm plots and does not disturb the existing land holding pattern. This idea was suggested, along with a pilot project, at Bakolori, but was not taken up. Examples of insensitive design include the discharge of drainage water onto a large area of valuable floodplain land outside the project area at Bakolori (see Adams, this volume), and the design of irrigation fields over existing settlements and under a large lake at Dadin Kowa.

The lack of any provision for livestock (both that held by people in the project area and by nomadic Fulani herders) has also been a problem. Dislocation of the existing communications network between villages, particularly for livestock and people on foot (and also those on bicycles and motor cycles who use the foot paths) as a result of the superimposed rectilinear pattern and the barrier created by canals and drains, has also caused serious disruption and antagonism. Health aspects have also not been considered in the design, with the result that bilharzia and malaria are likely to become serious problems. No provision has been made for village water supplies, although this could have been easily and cheaply done, with the result that villagers are now drawing water from canals and drains for domestic use as well as using them as latrines.

Difficulties have been experienced in getting farmers to participate in dry season irrigation, mostly due to the fact that many farmers are engaged in other, often more profitable, activities during the dry season. At Bakolori some people on dry season occupation could earn up to 350 Naira clear surplus in each dry season (Adams 1980), and unless irrigated farming can give this sort of return then unsurprisingly farmers will be reluctant to partake. Another difficulty with dry season irrigated farming is that land preparation for irrigation has to be carried out in October, while the staple food crop at Bakolori, late sorghum, is not harvested until December. In order to partake in dry season irrigated agriculture farmers have to forego their main food crop.

All these problems stem from a lack of awareness of the need for socio-economic considerations of project planning and development. Without a baseline social survey, no meaningful monitoring or evaluation of the projects can be, or has been carried out. What socio-economic factors were considered were based on superficial naive assumptions about the rural economy and showed little understanding of the role of dry season irrigation and off-farm activities, with resulting misconceptions about labour availability. The organisation of farm labour by household and the situation with regard to extended family communal working were not appreciated.

CONCLUSION

In summary, then, due to the high occupancy rates and the highly developed land holding pattern coupled with small plot sizes and the existing high level of agricultural activity, formal, large scale irrigation schemes in riverine areas in northern Nigeria are inappropriate and disruptive. Compensation costs, regardless of which land policy is followed, are high, although small in relation to the cost of construction, and demand sudden inputs of government cash which are often not available. This is not helped by the fact that compensation costs are not included in project feasibility studies.

Such projects have been undertaken with sufficient forward planning and lack of liason with the people they affect. Project planning has been dominated by narrow engineering criteria and insufficient thought has been given to the social and economic effects of such projects. Small-scale informal dry season irrigation is already well established in the many areas in northern Nigeria and a more appropriate irrigation policy would seem to be to assist this with relevant help. However, it would probably be far more cost effective, as recognized by the World Bank, to assist the well established rain-fed farming activities and to make them more efficient by supplying better inputs and improving infrastructure. However, great care needs to be taken to ensure that all farmers and not just the larger ones benefit from this.

Contrary to policy aims, the present large scale formal irrigation projects are actively forcing smaller farmers, who were previously feeding themselves, off the land, either to the urban areas or as paid agricultural labourers. The payment of compensation for improvements and Rights of Occupancy is probably assisting this process.

FOOTNOTE

Alan Bird was the Senior Surveyor and acting Team Leader for the Bakolori Land Tenure Survey 1979-81, Senior Surveyor for the Dadin Kowa Resettlement project 1981-82 and Team Leader for the Dadin Kowa Irrigation project Land Tenure Survey 1982-83.

REFERENCES

Adams, W.M. (1980) Data from Yardala and Yarkofoji, Unpublished manuscript.

Adams, W.M. (1981) The Effects of the Bakolori Dam on Downstream Areas - Problems and Solutions, unpublished manuscript, Department of Geography, University of Cambridge.

Adams, W.M. (1982) Managing to Irrigate in Nigeria? In H.G. Mensching and V. Haarmann (Eds.) "Problems of Management of Irrigated Land in Areas of Traditional and Modern Cultivation". Hamburg 1982, International Geographical Union.

Adams, W.M. (1983) Downstream Impact of River Control: Sokoto Valley, Nigeria, PhD, University of Cambridge.

Bird, A.C. (1981) The Bakolori Agricultural Project MRT Land Tenure

Survey: An account of a survey for land re-allocation purposes. In Igbozurike . U.M. (Ed.) Land Use and Conservation in Nigeria. University of Nigeria Press, Nssuka, 1981.

Bird, A.C. (in press) Farmer participation in the planning and implementation of Nigerian large scale water resources schemes: Bakolori Irrigation Project and Dadin Kowa Resettlement Project compared. Paper prepared for the seminar on Development Agricole et Participation Paysanne , held at the Universite de Paris; Pantheon Sorbonne, October 1983.

Impresit Bakolori (Nigeria) Ltd., (1974) Bakolori Project: First Phase of the Sokoto-Rima Basin Development, Final Report. Impresit and Nouvo Castoro, Milan and Rome.

MRT (1979) Report of Engineers Representative on the Consolidation of Plots in the Bakolori Agricultural Project". Bakolori Project, Site Paper No. 26. Birnin Tudu, Nigeria.

MRT (1980) Bakolori Project, Outstanding Issues, Part 2, Volume 1, Land Management at Bakolori, Cambridge.

MRT (1981) Report on Farm Re-allocation Design by the Field Design Team" Bakolori Project, Site paper No. 50. Birnin Tudu, Nigeria.

Wallace, C. (1979) Studies in a town on the Kano River Project. Centre for Social and Economic Research, Ahmadu Bello University, Zaria, Nigeria.

LABOUR CONSTRAINTS IN THE IMPLEMENTATION OF IRRIGATION

J.M. Siann

WIDENING THE SCOPE OF ENQUIRY

Civil engineers are usually assumed to play a crucial role in the practical implementation of irrigation projects. This popularly held view probably arises from the awareness of their unique understanding of the principles of engineering hydraulics and the design of water-retaining structures. Although these are important factors in the planning of water regulating schemes, their significance is sometimes over-emphasised in Third World countries where development decisions are strongly influenced by technocrats. In these countries, the technocrats enjoy a disproportionate amount of respect from the particular government planners who have to wrestle with the logistics of trying to promote increased food production. Significantly it is in the field of agricultural engineering where the view of the expert is so highly respected.

Not often recognised, however, is the narrowness of the field of enquiry of the engineer. It is highly unlikely that an engineer would have a grounding in ergonomics within the context of a tropical environment, or could appreciate the possible sociological impact of the use of unfamiliar farming techniques, and possibly would not even have a thorough understanding of local physical conditions which could influence straightforward engineering decisions. Invariably, when one examines engineering texts on irrigation, one is struck by the fact that case studies are concerned with highly ordered economies such as in the United States, or with evolved peasant economies such as in India and the Far East where irrigation has formed part of the familiar farming regime for centuries. In these economies, the example found of highly structured irrigation systems cannot be taken piecemeal as blueprints for countries unfamiliar with this disciplined system of water management.

There is therefore a clear need to temper the engineer's enthusiasm in advocating novel techniques of water management. This might best be done by considering a growing number of examples in Third World countries where sought-after social and economic objectives have not been realised. In this way it might be possible to avoid repeating the errors in planning which have been shown to lead to disappointing results and, in some instances outright failure.

As a starting point, it is worth considering Carruthers (1978) who identified an operational model which should be of interest to professional groups concerned with irrigation planning (Figure 12). This model provides a useful starting point in establishing the engineer's relationship to other interested groups. Engineers (and presumably all other water technologists) feature as only one of the many pressure groups involved in promoting irrigation projects. The model does not attempt to rank the order of importance of the various promoters, but it is arguable that many of the other pressure groups under-rate themselves in deferring to engineers in the making of

FIGURE 12

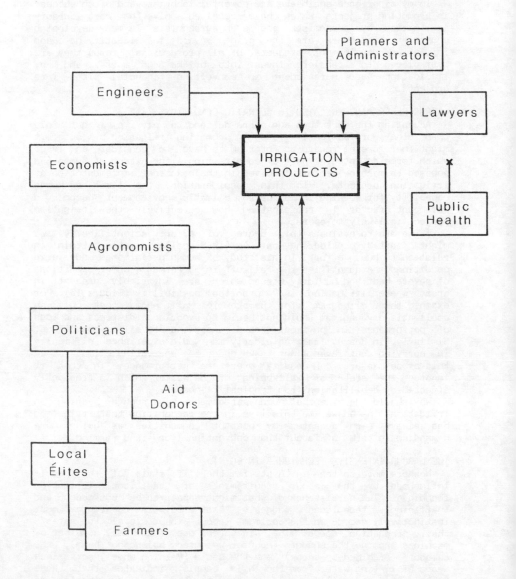

decisions on the viability of particular projects.

In my experience engineers are powerful lobbyists and often succeed in promoting projects which have received only the most cursory examination by economists and even agronomists. In many developing countries, engineers are the prime movers in identifying and initiating development projects. Of all professional groups they are best placed to translate ideas into bricks and mortar and, by implication are more adept in converting financial plans into projects.

LABOUR PERFORMANCE RELATED TO CLIMATIC CONDITIONS

Accepting the fact that engineers do tend to play a larger role than other professional groups in promoting irrigation projects, it might therefore be rewarding for them to have to consider an issue which forms an integral part of construction - the extent of human (as opposed to machine) effort involved in the practical implementation of irrigation projects. In this consideration, the limit of human capability for working in a harsh climatic environment becomes a paramount factor in anticipating how effectively the irrigation project is likely to operate.

If we are to evaluate this degree of effort scientifically, we might start by considering what Lee (1969) calls 'relative strain' in relation to human effort. In his study of human performance under arid conditions, he quantifies the fall off in performance under conditions of severe heat and humidity stress which are invariably present in areas where irrigation is a practical possibility (Figure 13). For example the 'climogram' for Bamako in Mali shows that only an acclimatised young man would not begin to experience distress and loss of performance in physical tasks under the prevailing conditions. Similarly, in Basra, Iraq, an elderly man would experience difficulty in carrying out even minor tasks under the enervating climate. Even this crude measure can easily overstate performance, in that it ignores the state of well-being of the worker which is frequently affected by debilitation from tropical illnesses.

Why should the extent of human performance have such a bearing on irrigation? The answer to this lies in the recognition that irrigation has always been accepted by peasant communities as being more demanding in terms of labour than conventional dry-land farming.

THE LABOUR-INTENSIVE TECHNIQUES IN EGYPT

Nowhere more so than in Egypt is the life-style of the people influenced by the labour requirements of traditional irrigation techniques. The persistence of what might appear to be out-moded and inefficient techniques suggests that Egyptian peasantry must instinctively decide on the optimum labour input level to satisfy their production needs. The widespread use of the very simplest of machines such as the shaduf (counter-weighted balancing beam), the tambour (Archimedes screw) and the saqia (Persian wheel) as various means of lifting water from the Nile canals, indicates that these techniques are still viable even with the ready availability of mechanical pumps.

By the use of these irrigation techniques, a system of water

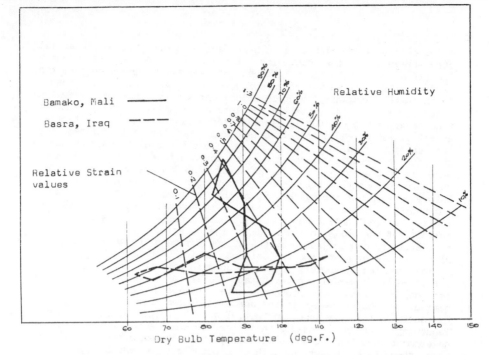

2A: CLIMAGRAM ON RELATIVE STRAIN CHART

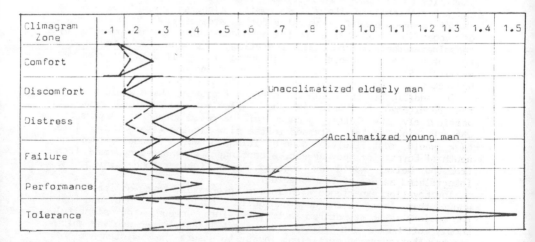

2B: SIGNIFICANCE OF RELATIVE STRAIN ON PERFORMANCE

FIGURE 13 : EFFECT OF CLIMATE ON PERFORMANCE

An estimate of the strain imposed on the body by a combination
of temperature and humidity (After Lee, 1963)

regulation has developed in Egypt which is mirrored in the social stratificaton of the community. The various inputs are closely identified with individual groups of workers commencing from the landless itinerant worker, proceeding to the tenant farmer (felaheen), to the small landowner living in close proximity to the irrigated lands, then to the absentee landlord and finally to the government functionaries are various levels each responsible with a particular aspect of water regulation. In this complex but logically defined way, the carefully balanced system of water use has evolved which ensures that no irreversible damage is done in the short-term. It is only when major irrigation projects like the Aswan High Dam are imposed on this delicate structure that repercussions are felt which endanger the future survival of the system. Since the construction of this large impounding reservoir, the absence of the normal seasonal floods and their 'flushing' action has resulted in progressive increase in the salinity of the soils in lower Egypt. The damaging effect of this loss of fertility can be judged by the steady stream of urban migration; a clear example of upsetting the delicate social/technology balance.

The ordering of society, which is so much a feature of Egypt, is sufficiently powerful to withstand political upheavals and land reforms, even those as profound as were initiated in the Nasserite period. No major change in the agrarian system related to irrigation has survived for very long. In fact, I have been told by dispossessed landowners that they consider their land as only being 'loaned' to the government and that it would ultimately be returned after the reforms were 'shown to be disastrous'. This reluctance to accept change is a by-product of the regulatory system which ensures the necessary stability to continue a socially adjusted system of irrigation. Thus characteristically, within the family clan, sons would operate the shaduf to produce sufficient water for an area of 2,000 to 3,000 sq.m of vegetable production; in the evenings this would be augmented by the older members of the family operating the tambour for a few hours to water the canal-side crop of berseem; and, if a larger area was available for crops, the family might use a saqia with a blind-folded bullock to irrigate an area as much as about two hectares.

In addition to the perennial need to irrigate, there would also be the requirement to maintain the inflowing and drainage canals, particularly the latter because of the rapid spread of papyrus and water hyacinth if left untended in the slow flowing waterpaths. This task would most likely be done by itinerant workers usually from the south of Egypt. For the harvest, a contingent of share-croppers (often from the Delta) would be engaged, thus further underlining the interrelated nature of task-sharing.

The Egyptian experience with labour intensive methods of irrigation shows that these techniques can quite readily compete provided that the society is so structured as to take into account the plentiful supply of human and animal labour. A lifestyle has evolved which ensures that there is sufficient labour to ensure the maintenance of the delicate hydrological balance required to safeguard the fertility of the farmlands. Not only has there to be adequate labour for raising the water into the canals, but the ever present danger of salinity and the progressive sterility of the soil necessitates a constant vigil over the rate of natural drainage.

Therefore by examining the example of Egyptian practices, one can begin to appreciate the extent of labour requirements which are required for irrigation as opposed to rainfed farming. Stern (1979) observes that with rain cultivation the farmer is able to undertake non-farming activities much more readily, but with irrigation water has to be applied without fail when it is due, and in the correct amounts. The farmer therefore becomes 'tied' to his land. Cantor (1967) further endorses this by stating that "irrigation does not only affect crops and crop yields but it also determines to a considerable extent, the social organisaion of the farming units". This is clearly the case in Egypt as it is, for example, in Java where 'irrigation societies' have been created consisting of those families owning ricelands irrigated from a single watercourse.

NIGERIA'S EXPERIENCE OF IRRIGATION

Turning to the as yet embryonic culture in Nigeria one attempts to seek further endorsement of the importance of manpower in the success of irrigation schemes, and finds the evidence already present. Whilst the use of irrigation is mainly confined to a few schemes in the North, there has in recent years been a strong interest in extending the practice to other areas of the country wherever prolonged water deficits are experienced.

If we examine two of the longest established schemes, in the Bacita Sugar Estates and the South Chad Projects, we are already aware of the disquieting evidence that Nigerian society has as yet not become attuned to the discipline imposed by an irrigation-assisted system of farming.

The Bacita Sugar Estates had by 1963 developed into an area of 3,000 ha of cane, irrigation being used to supplement rainfall during part of the wet season and to provide all water requirements in the dry season. Since their inception, the estates have relied on seasonal workers but with the widespread increase in the incidence of schistosomiasis, the scheme has become upopular and is now plagued by a severe labour shortage.

Similarly the second project, the South Chad Project, demonstrates a brave attempt to supplement the low rainfall in a promising area by irrigation so as to ensure more reliable yields of wheat and rice. The project is eventually intended to provide 100,000 ha of irrigated land by using the natural storage of Lake Chad. In this part of the country rainfall is the lowest and most unreliable in Nigeria, and as a result crop failures are frequent. It is for this reason that the scheme was proposed for the local inhabitants, but reports suggest that only incomers have to a large measure been attracted to the scheme. The cost of constructing the engineering works to bring water from the Chad, which has a large seasonal variation in level, has proved to be exorbitant and even at this stage the scheme has been variously described as "very risky" and "more a social venture than an economic proposition".

Thus Wells (1974) suggests that the success of irrigation projects in Northern Nigeria can only be judged within the context of a situation fraught by severe constraints. One of the basic preconditions for economic survival is the need for double-cropping so as to utilise irrigation schemes to their fullest extent. Whether there is sufficient labour to achieve this level of utilisation is an issue which does not appear to have been adequately considered before embarking on these projects.

Other examples in Nigeria further highlight the lack of initial appraisal or ignoring of advice given by independent specialists. May (1981) mentions the Bakolori Dam which has turned out to be the most expensive irrigation project in West Africa with a capital cost of £200m for a mere 70,000 acres of arable land (almost 3000 per acre). This waste of scarce resources has even been condemned by the locally based Institute of Agricultural Research at Zaria where the projected yield has been estimated at no more than £50 per ha for irrigated wheat.

NIGERIAN SCHEMES FOR RESETTLEMENT

Resettlement in "development" villages has been one of the ways used in Nigeria to assemble sufficient manpower to ensure the success of previously tried agricultural schemes. The first concerted efforts in trying to create new agricultural settlements were initiated in the Western State in the early 1960's. The Western Nigeria Farm Settlement Scheme, planned and inspired around the Israeli moshav concept was seen to promise major advantages compared with traditional villages and their falling agricultural output. With a new institutional form providing coherent land holdings, common processing and marketing and the possibility of introducing more capital intensive methods in a mixed farming environment, a way was seen to revolutionise the failing traditional system of bush-fallow crop raising.

The limited success of the Western Nigeria Farm Settlement Scheme can be judged by the fact that after 5 years only 1,500 families were involved in the programme and the settlements were already plagued by resignations and local attitudes which had become generally negative. A study of the reasons for resignation revealed that many of those recruited felt that they were not intending to become permanent members but had used the opportunity for obtaining govenment training. What is more, the intended demonstration effect to surrounding farmers so as to encourage them to join the scheme has remained unrealised.

Clearly the experience with the Western Nigeria Farm Settlement Scheme should serve as a pertinent reminder when considering the viability of future irrigation schemes. Any irrigation project involves the establishment of a structured settlement pattern. In the context of Nigerian society, this has not yet been shown to be an unqualified success. The possibility of producing enough incentives to entice a sufficient number of able-bodied settlers seems unlikely within the present constraints on land tenure. Even the recent moves to nationalise land under the Land Use Decree have apparently failed to overcome this problem.

Thus in Nigeria there appear to be three major constraints which have to be overcome in the course of establishing irrigation schemes. Firstly, there is the reluctance to abandon the traditional system of shifting agriculture. Second, there is the difficulty of acquiring land suitable for new settlements, and third, there are the difficulties of enticing a sufficient number of able-bodied people to participate. Added to these essentially sociological difficulties is the need to establish good marketing arrangements. The present well-established periodic markets already relate to the important major centres and it would be difficult to encourage their extension into new areas, where markets would be generally smaller and less lucrative.

If the benefits of irrigation schemes are not particularly self-evident, why is there a persistent and vocal force of opinion in favour of new projects? Williams (1980) argues dismissively that irrigation projects are promoted by businessmen who are intent on surplanting the peasant farmer and taking over their land for wage labour employment. With their access to capital funds they are better equipped to fund the inputs necessary for dry-season farming and have better connections with the urban markets. He cites the example of the Bakolori Scheme in Sokoto State where many settlers have already sold their irrigated farms and are now reduced to the status of wage-earners.

AN APPROPRIATE SIZE OF IRRIGATION SCHEME

Is there any way of determining what might be an appropriate size for an irrigation scheme? One would think that in the light of the earlier discussed requirement for high levels of manpower input into irrigation assisted farming, it might be more sensible to examine more appropriate methods for peasant farmers to boost their output without requiring them to move to re-settlement areas. The first precondition for irrigated farming which stems from the labour requirement, is the notion that irrigation is unlikely to be successful as a supplement to rainfed production unless the population has reached a certain critical density. In other words, irrigation is effective as a means of raising agricultural productivity only where land has become a major constraint on future development. Evaluating this critical density will naturally depend on local circumstances but it is clear that in all cases a substantial degree of co-operation must exist between farmers so that all labour requirements can be covered notably in the impounding, storing and distribution of conserved rainfall.

Ansell and Upton (1979), in their assessment of a viable size of settlement for irrigating from surface storage in southwest Nigeria, suggest a group of 56 families. In their view, a group of this size acting collectively could earn an adequate return to cover the servicing cost of an irrigation scheme based on a ten-year recovery period. Even with this optimal size, other criteria such as valley topography and the availability of a suitable impounding site are bound to have a critical influence on the viability of the scheme. Nevertheless, one is often struck by the extent of unrealised potential when one sees the abundance of suitable sites for small dams near inselbergs and on smooth outcropping granite lenses. Given that these criteria can be met, there remains the over-riding difficulty of servicing such a diffuse arrangement of settlements. The cost of providing the necessary level of infrastructure could prove to be unacceptably high.

FARMERS' CONCEPTION OF ECONOMIC RETURN

Apart from the factors which might affect the physical logistics of irrigation, there is also the attitude of individual farmers to what they consider to be economic behaviour. They might decide to adopt a strategy of return which provides them with a reasonable amount of leisure time and opportunities for other social pursuits. Or they might adopt a strategy which minimises the level of loss which might be incurred. Sir M. MacDonald and Partners (1951) cite an example in the central Jordan Rift Valley where there is a huge gap between the technological s kills which are readily available and those which are actually applied. The techniques which could lead to the more efficient use of water are spurned in favour of the traditional system of "wild-flooding" irrigation because it provides what is considered to be an adequate return.

IDENTIFYING THE PERTINENT OBJECTIVES

Clearly there is a need to relate irrigation techniques more closely to the extent of their likely adoption by farmers. More cognisance has to be given to the preferred lifestyle and other socio-economic determinants than the consideration of the bare economics of water distribution. Where in the past the emphasis has been on assessing the direct costs, i.e. the cost of impounding and channelling the water, in future there will have to be more attention on the return to the farmer in terms of crop production and disposable surplus value.

The types of analysis most frequently used hitherto appear to be deficient in the aspect of labour input requirements. This has resulted in many unfulfilled objectives. In a majority of instances the farmer is seen only as an operative and little consideration is given to what he might perceive as economic behaviour in the light of his experience of local conditions. If there is to be any future for irrigation in an area not already conversant with it, the problem of labour inputs will surely have to be addressed.

REFERENCES

Ansell, A. and Upton, M. (1979) Small Scale Water Storage and Irrigation. An economic assessment for South West Nigeria, University of Reading Development Study No. 17

Cantor, L.M. (1967) The World Geography of Irrigation, Oliver and Boyd, Edinburgh, 1967

Carruthers,I.D. (1978) In Widstrand C. (Ed.). Water and Society: Conflicts in Development. 1. The Social and Ecological Effects of Water Development in Developing Countries, Pergamon Press.

Lee, E.H.K. (1969) Variability of Human Response to Arid Environments in Arid Lands in Perspective,(Ed.) G. McGinnies and Bram J. Goldman, University of Arizona Press, Arizona.

MacDonald & Partners, Sir M., (1951) <u>Report on the Proposed Extension of Irrigation in the Jordan Valley</u>, Westminster, Cook Hammond & Kell.

May, B. <u>The Third World Calamity</u>, Routledge Kegan & Paul, London, 1981.

Stern, P. (1979) <u>Small Scale Irrigation</u>, Intermediate Technology Publications, London.

Wells, J.C. (1979) <u>Agricultural Policy and Economic Growth in Nigeria</u>. Ibadan, Oxford University Press.

Williams, G. (1980) <u>State and Society in Nigeria</u>, Afrografika Publishers, Idanre, Ondo State, Nigeria.

MISMANAGING THE PEASANTS: SOME ORIGINS OF LOW PRODUCTIVITY ON IRRIGATION SCHEMES IN THE NORTH OF NIGERIA

R.W. Palmer-Jones

INTRODUCTION

Smallholder irrigation projects in Africa are, with a few exceptions but in common with those in other parts of the third world, notorious for their low productivity. African peasants have been reluctant participants in these government sponsored irrigation projects: they have refused to commit their labour, to finance the purchase of tractor services, seeds and fertilisers, to repay credit, to pay for land "improvements" and water, to turn up for cultivations, canal maintenance, watering and so on, on time, to grow required crops and to sell their crops to the authorised agents, or even to allow their land to be "developed for irrigation".

There have been two explanations for this phenomenon. There are those who claim that the peasants are lazy, stupid and greedy, as evidenced by their failure to take advantage of the attractive opportunities presented to them: and there are those who argue ultimately that the incentives are not in fact enough in view of the actual availability of the required resources and the more attractive uses for them open to the peasants, and to which they prefer to divert their time, finance and land.

Supporters of the first argument argue that coercion is required in order to get peasants to do what is patently in their own interests: they point to "successful" schemes such as Mwea and Gezira in which coercion, in the form of tenant discipline and management control guaranteed by revokable tenancies, is prominently involved.

Supporters of the alternative view point to the low and risky returns to participation, the low yields of crops and productivity of inputs, the large amounts of labour involved for low and unreliable returns, the unreliability of supplies of water, fertilizer, seeds, tractor hire services, credit and of access to markets for products: and to the importance of other farming activities, off-farm incomes, home production, socially necessary activities and so on. It is argued that coercion is irrelevant and that its use is a device to exploit the peasants for the benefit of the projects' sponsors in Government, the local elite, engineering contractors and consultants, agribusiness firms and so on. Participation and consultation are required, rather than control, coercion and discipline, so that farmers may become aware of what is involved and feel committed to the project: that they may exert some control over what is done so that it more closely accords with their needs and resources. Also they question the success of projects such as Mwea and Gezira.

DISTRIBUTION AND BENEFIT

In this paper, based on the example of the Kano River Project (KRP) in the north of Nigeria, I shall argue that neither view is totally right, or rather that there are elements of truth in both views. On

96

the one hand yields and productivity are often low, and attractive alternatives are available. However, on the other a factor contributing to low productivity is the lack of coordination among farmers which contributes to inefficiency in tractor operations and water distribution and it is not necessary that productivity should be so low. Productivity remains low, however, since the management not only fails to solve the problems of coordination, but exacerbates them by using its control of crucial resources, especially tractors, in a way that creates disincentives for farmers. At the same time this is advantageous for management and staff who benefit personally from their control by extorting bribes and access to favourably situated irrigable plots, and preferential supplies of necessary inputs like tractors, water, seeds and fertilizers for themselves and their clients or patrons. This is achieved by threats and bribes, the denial or offer of favourable access to inputs, resolution of disputes and other valuable or harmful services at staff disposal – to landholding farmers.

This personalised system of distribution further militates against concerted farmer action, since the same combinations of threats and bribes can be used against those who might try to organize collective action against corrupt or inefficient services. Moreover this pattern of personal gain from bureaucratic position and the biased distribution of state-controlled goods conforms to the political system both under military and civilian governments. While preponderant economic power is concentrated in the state, economic gain has been conditioned by access to political power and influence, either through connections with the bureaucracy or through political (military or elected) position. Economic gain is used to bolster these affiliations and positions (as well as personal accumulation and conspicuous consumption). A number of devices are used to maintain control over state resources including direct physical repression, cooptation, the organization of subservient institutions, and the manipulation of factional divisions. Since the state functions not in the general interest but in response to powerful groups and pressures, these mechanisms reproduce and serve the interest of those who control the state. The main source of accumulation is access to the State, rather than, say, control over wage labour, because of the preponderant economic power of the State, and State activity is oriented primarily to the distribution of revenues and other benefits, rather than production. It is access to timely and cheap supplies, controlled and rationed by the State, of fertilizers, tractors, water, combine harvesters and so on that ensures private profits rather than efficient organisation of production, especially if the credit given for supplies need not be repaid.

Even where such rampant use of the State for private benefit does not occur, the general role of the State in extracting surpluses, combined with the inevitable start-up problems of irrigation schemes in these environments, leaves few resources and too little local control for the efficient solution to the externality and public goods problems of smallholder irrigation schemes. These problems can be seen as the lack of suitable institutions to allocate resources under conditions of market failure. Through an analysis of the form of

97

market failure it may be possible to provide a basis for a more productive discussion of irrigation schemes than the counter-position of the two views caricatured above. Incentives and coordination are both required.

LOW YIELDS, RETURNS AND INCENTIVES

An account of the development of the 'peasant as obstacle' view apropos of irrigation projects in the north of Nigeria can be found elsewhere (Palmer-Jones, 1980, 1981, in press) and I will not rehearse the evidence here.

The opposing view, that incentives and resources are inadequate for the majority of farmers is, almost of necessity, based on less than wholly satisfactory evidence. Furthermore it is doubtful if any evidence could convince those most fervently committed to the value of irrigation since their argument is based on potentials, rather than actual performance. Thus the Hadejia Jema'are River Basin Development Authority (HJRBDA) point to the highest yields and returns obtained and dismiss arguments based on averages, risk, or the lower yielding plots. A handout given to visitors to the Kano River Project (KRP) in 1982 gives a wheat yield of 5 tonnes per hectare. This brochure also shows average net incomes of 1,500 Naira (N) per ha; since the wheat price was around 300 Naira per tonne this entails the claim that average yields were 5 tonnes per hectare. (1)

The scheme was initiated in the early 1970's with a pilot project at Kadawa. It aimed to introduce a dry season irrigated crop in an area already fully cultivated by peasants in the summer rainy season; traditional crops are in the ground from May to November. It is a surface gravity scheme on upland soils; in the early years the water supply was pumped into the canal system. The land is developed for irrigation and plots are handed back to the original farmers with minor modifications to boundaries to conform with the irrigation layout. The main irrigated crops have been wheat and tomatoes which are sold generally on the open market. Because of the rotational requirements which prevent tomatoes being continuously grown, and the thin market for fresh tomatoes, wheat is grown on most of the scheme; wheat production to substitute for rapidly rising imports is a major objective for the project and its success is seen to be mainly dependent on wheat.

The basis of the HJRBDA yield figure is the correct report of the yield of a few farmers, typically those with whom extension staff have frequent contact, of the number of bags of threshed wheat they obtained divided by the area recorded in scheme records. No system of scheme-wide sampling and formal measurement is conducted, nor has been since the scheme started, and since nearly all produce is sold privately no indirect method is available. Thus there is no reliable record of the areas of crops cultivated or yields kept on the scheme. This, combined with the patently unreliable figures given by the scheme, e.g. in such handouts as that mentioned above, makes empirical assessment of the scheme difficult.

The only evidence available at KRP comes fom a multiple-visit farm economic survey I conducted in 1975-77. (2) The sample of 30 farmers was representative of the population resident in the village whose lands were most affected by the pilot project; it obviously became less representative as the scheme expanded.

A very different picture of yields was given by this survey. Wheat yields were 1.6, 3.1 and 2.5 t/ha in 1975-6, 76-7 and 80-1 respectively, ranging from 0-3.6, 2.42-4.5 and 1-4 t/ha. (3) The improved average yield in 1976-7 is due both to the abandonment or giving out for cultivation by non-sample farmers (4) of low yielding fields, (5) and to the more favourable weather in the latter year. (6) The data for 1980/1 have to be treated with considerable caution; but they reflect not only reasonably favourable weather but also an increasing ability of these farmers to choose to plant in plots which will give reasonable yields, rather than any drastic improvement in productivity on the scheme, since it was still reported that a major portion of the scheme was planted after Christmas.

Average incomes net of cash expenditure (7) from wheat of those households who cultivated wheat was 84 Naira ha in 1975-6 and 463 Naira (N) in 1976-7 when the average price of wheat at harvest was N 190 and N 270 respectively. (8) In 1975-6 30% of farmers made cash losses. Since 1976-7, cash costs have increased as has the price of wheat, but without more detailed information, net incomes from wheat in 1980/1 cannot be estimated. Nevertheless, given the small rise in wheat prices there is no way net average incomes from wheat could have reached N 1,500 by then. The main cause of the increase in income in 1976-7 was the increase in yield and harvest price. The rise in harvest price combined with the securing at last of a more reliable source of water (9) led to somewhat increased enthusiasm for wheat in 1977-8, and must account in part for the increase in wheat area cultivated by the sample over time. But this also reflects growing wealth of farmers mainly from off-farm activities, and, for some, successful accumulation through irrigation, and, perhaps, a greater ability to work within the constraints of the scheme.

The story for tomatoes, the only other major irrigated crop, is similar in that actual average net incomes, although higher than for wheat, seem to fall well short of those claimed by the scheme in 1980/1 and earlier. Actual yields averaged around 4 t/ha, while net incomes in 1976-7 were N 1,900 ha .

The scheme was also supposed to raise yields of rainy season crops such as maize, millet and groundnuts. With the partial exception of maize these crops fare no better on the scheme than outside (for groundnuts see Palmer-Jones 1978). Maize has proved an attractive crop for a minority of farmers (3 in the sample in 1976) who could obtain access to tractors, improved seeds and inorganic fertilizers, and who could rely on producing enough millet and sorghum for their own subsistence or had a sufficient and reliable off-farm income to purchase staple food.

ATTRACTIVE ALTERNATIVES

Farmers needed to supply time, effort, management skills, and finance; these were required at specific times of year and places to fit in with the farming seasons. Finance was typically required at harvest time when prices were at seasonal lows; prices could as much as double or more in the two months following harvest with consequent implications for the opportunity cost of crops which would have to be sold to finance (relatively) unprofitable irrigated crops (Palmer-Jones 1977). Also such finance had alternative uses in the numerous trading and crop-storing-against-a-price-rise possibilities available. In that subsistence production was harmed by the scheme prohibiting the growing of guinea-corn, (10) some extra finance would be required for food purchases by some households.

Labour, encompassing time, effort, and management, was typically required when it did have alternative uses especially once the scheme was established and had generated considerable direct and indirect demand for labour and services. For the period from November 1975 to April 1977 off-farm cash income comprised on average 51% of net household income. (11) For 15 out of 30 households government paid wage labour provided the main source of off-farm income (4 of these were migrants), and for 5 others casual work for the Government provided some income. The bulk of off-farm cash income (56%) of the average household was obtained by full-time wage labour for the scheme either on the government farm or construction. At the time the minimum government wage rose from N 1.50 per day to N 2.00 per day; for full-time employment this gave an annual cash income of between N 552 and N 720, which was greater than the sum of average wet and dry season net farm incomes. Indeed the income from a government wage job exceeded the total net farm income of 70% of the sample.

Of those households who received off-farm incomes of less than N 100 in the survey period at least one was lying; three were agricultural labourers unable to get other more highly paid work (two were simple, one a maguzawa (12)) three were old and in fact received income supplements from their sons, and one was a religious adept, who probably received extra income in kind not recorded. Sixteen out of the sampled 30 obtained off-farm incomes greater than N 500 in the 18 months of the survey, mainly from wage labour for the government by themselves (9) or members of their household (3), and from trade (4) in cattle, kola nuts, retailing manufactured goods, and imported foods.

The full story of the labour time required for off-farm activities, farming (wet and dry season), gathering of natural produce, home building and maintenance, water and firewood collection, household purchases, animal tending, domestic labour, and necessary social activities escapes the figures collected; but on average over 3,600 hours were recorded as spent by household members in off-farm income-earning activities in the 18 month survey period (over 2,200 by the household head alone - or 122 hours per month). The time spent was fairly evenly spread throughout the year with the average household spending 72 hours in the month with the lowest total. Another 1,160 hours of family labour (65% - 760 hours - by the household head) were recorded as spent in farming in the 18 month survey period; this too only tells a part of the story since much time waiting for irrigation water, trying to obtain fertilizers, seed, tractor services, and supervising them, as well as acquiring knowledge of how to work with the scheme and to adapt to its, somewhat rude, interevention were necessarily under-recorded.

Thus for the household head alone the recorded activities accounted for 164 hours per month or more than six hours per day. This is clearly an underestimate since labour on household production was not included, as well as the points noted above. It would take too long to go into all the detail of the household labour available since there were a number of joint households with different implications for the availability and control of labour of men, women and children.

But it is evident that the household head was not grossly under-occupied. What evidence I have suggests that there were good returns to be had from activities other than farming, (13) and that failure of peasant supplied inputs could partly be explained in this way.

CIRCULAR CAUSATION?

Nevertheless the question remains "why, if technically it was possible to achieve higher productivity in wheat production, with consequent implications for the attractiveness of participation, was higher productivity not more widely achieved"? Higher yield was the main determinant of higher net returns within the sample, and between years. The question then becomes what are the determinants of yield? Again cross-section survey data cannot be unambiguously interpreted; nevertheless it is clear that nitrogen fertilizer level and date of planting were most important (see Palmer-Jones 1980). This is supported by experimental evidence which also shows that yields decline steadily if planting is performed towards the end of December and into January. The effect varies with the year, being greater in some (1975-6) than in others (1976-7), (14) but can be expected more often than not (AERLS 1979). Also the response to nitrogen tends to decline with later planting.

On the scheme, planting of more than half the area has generally taken place after Christmas and this must contribute most to low productivity. The question then becomes, why was planting late? The answer to this question is disputed between those who hold the two views given above. Those who blame the farmer argue that farmers delay paying for tractor services in advance, and plant guinea-corn in their plots, forcing the scheme to delay cultivations; credit cannot be given because of the difficulties of recovering payment given that there is no control of marketing of the product, or of land.

Those who blame the inadequate incentives point to the lack of control by the individual farmer of the date of cultivations, because when a field is cultivated depends on when most of those with plots in it have paid, as well as the vagaries of the mechanical department. They also point to the previous unreliability of water supplies, to the sometimes low prices for wheat (there was no effective guaranteed minimum price), and to corruption and delays in input supplies on the part of the scheme. All this causes farmers to expect low returns and even cash losses, and dependence on factors beyond their control; (15) this may explain their reluctance, in view of the alternatives, to commit land, labour and finance to irrigation, especially since there was no reliable, speedy and unbiased procedure for resolving disputes either with other farmers, or with the scheme.

Where expected returns are higher, as in tomato production, there was no reluctance to pay for inputs in advance, but this still did not result in trouble free planting, and universally high levels of returns.

However, this does not appear to be the whole story. Farmers themselves claimed that some other farmers did obstruct the management of the scheme, and it is clear that there is a certain circularity in both arguments. There was a sense in which the problems were due to mutual distrust, and to the absence of a means of resolving disputes. Typically a complaint by a farmer, such as that land had been inadequately ploughed or levelled, or that water did not reach his plot because of the abstractions of others or poor canal alignment, could not be resolved because no one would accept the story told by the other side. Furthermore, once started, there was strong circular chain of causation whereby problems inadequately resolved lead to resistance, obstruction, and 'vandalism', often harming third parties who themselves then felt agrieved, and could not get their grievance resolved, in part because other farmers contributed to it, and so became obstructive, and so on. This vicious circle is given a further twist when it becomes possible for managers to benefit from these events, by taking land and bribes in order to assist in the resolution of disputes and provision of inputs. Complaints of corruption, and of managers of the scheme and powerful people taking preferential advantage of the land appeared to me to have increased by 1982.

Farmers were further alienated by explicit threats by the scheme to take away the land of those who did not cultivate it, or pay for (or repay) inputs. The management frequently attempted to institute a system of licenses for the land on the scheme. Such a solution, tenancies on irrigated land which can be revoked if tenants do not comply with the requirements of the scheme (or some other means of enforcing discipline), is widely advocated. It is also in some ways a theoretically attractive solution to these problems, especially if linked to a dispute and grievance arbitration procedure such as a management committee on which tenants are represented, and which is independant of management. However, not only is the management committee usually either neglected, or is hardly independent, but once the history of land tenure and its administration is known, it is not surprising that this type of proposal only served to reinforce farmers suspicions and non-cooperation.

THE LOGIC OF LAND CONTROL

On irrigation projects rights to land are dependent on rights to water and other inputs necessary to make productive and rewarding use of the land; tractors and fertilizer are apparently particularly important at KRP. Rights to land can be meaningless unless rights to water and tractors are assured; but when they are considered these rights turn into obligations of the landowner to provide his land, labour and finance with no guarantee of the provision of these necessary inputs, or assurance of return. Indeed those charged with provision of these inputs use them to extort resources from the farmers, drawing on traditions going back to pre-colonial practises. The Nigerian elite has little incentive to reform the system since the present practice suits them very well.

Changes in land tenure on irrigation schemes have been aimed at establishing control over the land by the managers of the scheme. Two reasons have been advanced; one, the first, was to extract the "economic rent" as a tax both to pay for developments and operation and maintenance and to "motivate" farmers to take up and make efficient use of the developed land. The concept of rent is inappropriate for peasant societies since markets do not adequately define costs and returns for other inputs. The question of motivation depends on ideas reminiscent of the 'backward bending supply curve of labour', if not the 'lazy native'. Both these arguments have been made redundant by greater understanding of peasant rationality and the imperfect markets they face.

MARKET FAILURE
There is an important confusion in the latter reason, in that no recognition is given to the structural obstacles to co-ordination posed by rational behaviour. It could be argued that the idea was to train farmers to co-operate by showing them the benefits of doing so; but this is to misunderstand the basis of the need for co-ordination. This, as I have argued recently, can be understood in terms of the many important externalities and public goods problems involved in irrigation projects (Palmer-Jones 1982). Thus top-end water users taking water when they please, generate an externality for bottom-enders by reducing the amount and timing of water available. Similarly farmers who grow wet-season guinea corn in an irrigation field at Kadawa, or who do not pay in advance for tractors, delay these operations (16) and impose a cost on others with plots in the same field. If there is no way those affected can get those imposing the externality to comply, and if it is in the private interest of the latter to act in the way that imposes the externality, there may be a conflict between collective and private interest.

The point is that unless an institution is established which will ensure that this conflict of private and collective rationality is resolved, then no training will induce rational individuals to co-ordinate when it conflicts with private interest, even if by acting selfishly the project may become so unproductive that it becomes non-viable and ceases to function, even for top-enders. Thus it is not merely a question of showing farmers the benefit, or training them to behave 'as if' they were co-operative, or socially rational; but of establishing institutions to resolve conflicts and modify private rationality to conform with social rationality by confering benefits and imposing costs.

There are a number of theoretical solutions to this type of problem which can, within the context of a market economy, be understood as a case of market failure. Solutions may range from the establishment of markets by, for example, establishing property rights in water, or by encouragement of private tractor hire services, through to coercive solutions such as revokable land tenancies, or turning the farmers into wage labourers on large-scale farms. The tenancy soluton is essentially the specification of labour (and other) inputs in a land contract. Bureaucratic administration backed up by legal sanctions is a particularly common solution.

Government-sponsored large-scale irrigation projects for small farmers undoubtedly entail many characteristics of market failure; (17) perhaps the two most important are, first, the externalities between farmers over, for example, taking water on a rota, or contributing to tractor costs on time and not growing guinea corn; and second, the elicitation of effort and honesty from the management in the performance of their work. Another 'market failure' is the difficulty of small farmers in insuring against risks of crop failures or economic losses.

It should be obvious why those constitute failures. Individual farmers have no control over the actions of others and this being so there is no reason why they should take them into account. Even though they would benefit from co-operation (in for example a water rota, or payments for tractors) once a plan of co-ordination was reached each individual would have an incentive to take water out of turn if it suited him whether he thought others would or not, and the plan would have to be monitored and enforced. Farmers do not have an enforceable property right in water, or contracts for tractor services specifying their timing and quality, to use as a basis for contracting with the irrigation management. Such rights are difficult to establish because of the externalities involved.

BUREAUCRATIC FAILURE

Market failure could not be such a problem if there were a reliable bureaucracy to allocate the inputs in an impartial way. Herein lies the rationale for a water management bureaucracy (or its reform) impartial between farmers, rationally and incorruptibly executing its functions. The efficiency of the bureaucracy, however, is itself public good and may depend on enforcement by a supervisory body, or an ethos, a bureaucratic loyalty or protestant ethic. It lacks any economic incentive to perform its functions. Even if it (the bureaucracy) were paid a portion of the results, the problem of teamwork would limit the effect of such collective incentives on individual performance.

Moreover the nature of the task, to allocate a valuable resource, adds the oportunity for private gain to the difficulty of supervision (18). This too results from the poor definition of property rights and consequent market failure. Thus while peasants do not have an enforceable right to water, nor do bureaucrats. Peasants can neither force the bureaucracy to provide contracted water, nor can bureaucrats sell water (or peasants buy it) in an open market. These problems are of course particularly acute where water supplies are unreliable, making water rights even more difficult to define and enforce, and providing bureaucrats with both covers for shirking, and opportunities for extortion. This follows because water flows, or tractor availability, become unpredictable and inevitably subject to the discretion of management; this discretion can be bought in a surrogate market, which may be far from perfect because of the extent of its 'illegality'. Nevertheless scheme managers, inadequately disciplined themselves, have incentives to destabilise the water supply since it enhances their discretion.

Finally it needs to be pointed out that the provision of a

collective good, such as a water rota, or ploughing schedule, by making people better off, provides a 'rent' that can be captured in part at least by the bureaucracy and, as Wade (1982) argues, passes upwards and redistributed through the bureaucratic and political system.

As a consequence of all this, the almost necessary market failure on public irrigation schemes is replicated by the tendency of alternative solutions, particularly the bureaucratic management of schemes, to fail. But why does the bureaucracy fail? Or does it, and is the problem not one of irrational peasants? After all, many irrigation scheme managers will claim that they are not corrupt and that they are efficient. And since they control access to information about the scheme, it is often difficult to contradict them.

I hope that sufficient evidence has been provided to justify the belief that the bureaucracy does, and is likely to fail. Researchers discovering that all is not well articulate the grievances of the farmers, but if they fail to understand the structural and historical dimensions to the problem they are likely to perpetuate the circle of misunderstandings and confusions. However there appear to be no easy solutions. Both the choice of solution and the sources of its failure are deeply rooted in modern Nigerian society and its recent history. The political conditions for the efficient and just administration of irrigation schemes may not exist, and in analysing experience and advocating solutions at least we can recognise this.

ENDNOTES

1. Evidence reporting farmers' claims of much less were dismissed as efforts to delude gullible expatriates, the complaints of a minority of delinquents and so on (see further below). Also it seems wrong to claim N1,500 as the net rather than the gross income.

2. At the Bakolori scheme a planning unit did collect information on crop yields and areas. Yields of wheat averaged over three years were 1.25 t/ha (42% of potential); tomato yields were 6.2 t/ha (34% of potential); 38% of the area developed and ready for irrigation was fallow in 1980-81 dry season (Etuk and Abalu, 1982). For other accounts of KPR see Baba 1974, Dotteridge 1978, Wallace 1980 and Jackson 1978.

3. Yields in the first two years were estimated by counting the number of bags harvested on each plot and weighing 1-5 of them, and dividing by the measured area planted and harvested (no allowance for canals, drainage areas). For 1980-81 the yields were estimated, from the same sample, by asking farmers in February 1982 the areas and numbers of bags for the 1980-81 season.

4. In many cases these were government employees such as extension staff and scheme managers. These could obtain higher productivity than local farmers as described below.

5.   Thus in 1975-6 20 (out of 30) farmers grew wheat on a total of 15.9 ha; in 1976-7 9 farmers grew 8.2 ha of wheat; in 1980-1 22 farmers grew 3 ha of tomatoes of which 2.5 ha was on land not cultivated by the sample farmers in 1975-6. Hence of the area cultivated by the sample in 1975-6 7.7 ha were not cultivated by sample of farmers in 1976-7. These 7.7 ha gave an average yield of 0.6 t/ha in 1975-6, considerably below the average.

6.   As pointed out below a major cause of low productivity is late planting in "normal" years. Late planting causes low yields when the average temperatures rise above 33 °C within 6 to 7 weeks of germination, i.e. for wheat planted after Christmas if temperatures rise in the middle of February the average yields can be expected to decline drastically. Long-term temperatures and data indicate that average temperatures rise to 34° C in the 3rd week of February.

7.   Cash costs were obtained from farmers interviewed at least twice weekly. They were checked with official prices and levels of inputs. Inputs obtained on credit from the scheme, all of which was repaid, were included.

8.   Gross incomes were N211 and N669, while cash and kind costs were N115 and N205; N30 and N94 for hired labour respectively (no record of payments to family labour, which were observed occasionally, are included), the rest is for tractor hire, seeds, fertilizer and water.

9.   The completion of the main canal which was fed by gravity from a lake removed dependence on pumps which farmers had learnt were unreliable.

10.   The scheme discouraged guinea corn production because it conflicted with the need to clear the land in October and November for preparation for irrigated crop production.

11.   Categories of off-farm income included the wage labour and cash income earning activities of adult males and females as reported by the household head. Obviously considerable errors are involved; of likely particular importance are income from produce trading and money or commodity loans.

12.   A non-Islamised group; they were at the time discriminated against in employment by the government in favour of Muslims.

13.   Even if the data were more satisfactory, the methodological problems of using cross-section data to estimate the marginal value of time in irrigation farming as opposed to other activities seem to me to make such an enterprise useless. These are partly the problems of the time-specific nature of work in agriculture, and of the risks of agricultural production, not to mention the recent rapid changes in economic opportunities which, given adjustment lags and costs, must make the survey period one of disequilibrium.

14. Another aspect of the improved yield performance of wheat in this year was the abandonment of fields that were likely to be late planted.

15. Another factor is the high labour requirement before planting to level land every year for irrigation.

16. Or makes them more expensive because of the suboptimal working conditions imposed by working small plots or requiring more equipment to undertake the same work in the same time, or by shortening the time available.

17. There are many apparently successful 'indigenous' irrigation schemes which show that failure can be overcome without government involvement, although of course this does not mean they always are.

18. The physically dispersed nature of agricultural production and its effects on supervision costs must not be forgotten.

REFERENCES

AERLS (1979) Report on Wheat Production and Marketing in Nigeria. Agricultural Extension Research Liaison Services, Ahmadu Bello University, Zaria.

Baba, J.M. (1974) Induced agricultural change in a densely populated district: a study of the existing agricultural system of Kura district and the projected impact of the Kano River Project. Unpublished Ph.D. thesis, Dept. of Geography, Ahmadu Bello University, Zaria.

Dottridge, M. (1978) Aspects of social and economic development in Southern Kura District before implementation of the Kano River Irrigation Scheme. CSER, Zaria, 1978.

Etuk, E.G. and G.O.I. Abalu, (1982) River basin developmnt in Northern Nigeria: a case study of the Bakolori Project. International Commission on Irrigation and Drainage (ICID) Proceedings of the 4th Afro-Asian regional conference, Lagos, 1982.

Jackson, C. (1978) Hausa Women on Strike. Review of African Political Economy 13: 21-36

Palmer-Jones, R.W. (1981) How not to learn from pilot irrigation projects: the Nigerian experience. Water Supply and Management 5: 81-105.

Palmer-Jones R.W. (1977) A Role for Credit on the Kano River Project. Mimeo, Department of Agricultural Economics and Rural Sociology, ABU, Zaria.

Palmer-Jones, R.W. (1978) The Role of Groundnuts in Large Scale

Irrigation Projects.   Paper presented at the Conference on Groundnut
Production in Nigeria, Lake Bagauda Hotel, Kano, 1978.

Palmer-Jones, R.W. (1980) Why Irrigate in the North of Nigeria.
Paper presented at Seminar on Change in Rural Hausaland, Kano,
February 1980.

Palmer-Jones R.W. (1982)   The Elementary Economics of Irrigation.
Paper presented at conference on Water and Development in South Asia
of the Institute of British Geographers, Cambridge, Nov. 1982.

Palmer-Jones, R.W. (in press) Irrigation and Agricultural Development
in Nigeria, to be published in Watts, M.J., (ed.) University of
Berkeley Press, 1984. (forthcoming)

Wade, R. (1982)   The System of Administration Corruption; Canal
Irrigation in South India, J. Dev. St. 18(3).

Wallace, T. (1980) Agricultural Projects in Northern Nigeria. Review
of African Political Economy.   17: 60-70.

THE USE OF REPERTORY GRID ANALYSIS IN THE STUDY OF FARMERS' CHOICE OF CROPS

John Briggs

INTRODUCTION

A sense of frustration has emerged over the last decade in Africa at the apparent ineffectiveness of development strategies in stimulating and sustaining economic and social progress on that continent. Undoubtedly, many factors militating against the development process are outside the control of African peoples and governments; such might include low and fluctuating commodity prices on world markets, high fuel prices and the current world recession. However, during the 1970s, increased attention has focussed on what some perceive to be the inappropriate nature of development strategies within African countries themselves; a factor over which they can exercise a considerable degree of control. During the late 1960s and early 1970s, the emphasis of development in Africa shifted from the urban/industrial to the rural/agricultural sector, although, clearly, some countries, like Tanzania, went further than others in this respect. Nevertheless, although the direction had altered, the approach remained largely unchanged in the sense that development was still seen essentially as an institutional, administrative problem with development schemes and plans being imposed on rural people, whether wanted in that particular form or not (Lele 1975; World Bank 1981). Clearly, a development approach of this type is markedly centralist and makes the assumption that local population activities respond only to externally-initiated change (Long 1977).

An alternative approach requires making greater and more effective use of rural people's knowledge and experience, identifying their priorities and motivations, and incorporating local inhabitants into the pre-planning and implementational stages of development schemes and plans. Development projects should proceed with the people to be affected, rather than for them, an approach which may be termed development from below (Stohr and Taylor 1981). An increasing number of observers have called for this type of approach, so as to match planning objectives with peasant needs and aspirations, which in turn necessitates careful, micro-level studies to be undertaken (Chapman 1974; Floyd 1977; Barker 1979; Stohr and Taylor 1981). Development has to be more than a mere technocratic process, planned and implemented impersonally by, among others, "rural tourists", this particular breed being planners and administrators with only limited contact with those for whom they are planning (Chambers 1983).

The rationale for the present study evolved out of the theoretical points outlined very briefly above, and, in particular, the limited achievements in socio-economic progress in Africa thus far, and the fund of ecological and economic knowledge possessed by Africa's rural inhabitants. The problem remained, however, how to identify and use this knowledge. The focus of the study is on one small area of peasant knowledge and motivation, that of crop choice, identifying reasons underlying these choices. A methodology was required which would satify two criteria; firstly, the reduction of researcher interference in the understanding of crop choice, especially relevant in a cross-cultural situation; and secondly, the provision of an insight into how farmers choose their crops rather than how the researcher thinks farmers choose their crops.

The rest of the paper is divided into three sections. The first briefly describes the relevant agricultural characteristics of the chosen study area in central Sudan. The following section describes the operation of the methodology used. The final section provides a brief critique of the methodology, both its theoretical and practical aspects.

THE STUDY AREA

The area selected for study is centered on the Khogalab irrigation scheme, located about 35 km north of Khartoum on the east bank of the main Nile. The irrigation scheme is just over 1,000 feddans in area (one feddan is 0.42 ha) of which 924 feddans are cultivated (the remaining area being put over to access roads and canals), and is farmed by 231 farmers. Although the scheme is relatively large by main Nile standards, it is nevertheless very typical of irrigation schemes in this part of Sudan. It is ecologically constrained by the desert margins some two miles to the east of the river. It was established during the cotton boom of the mid-1950s. It is not under centralised control by the state (unlike many schemes on the White and Blue Niles), and is managed by an elected committee of local people.

This last point is important in that the committee are answerable to the scheme's members; moreover, the degree of management is minimised, concerning itself with maintaining and administering the use of the canals, pumps, the scheme's four tractors and various other types of equipment. Payments for hiring a tractor are made at the end of the year, in an attempt to reduce scheme-members' needs to obtain credit, and, in addition, members are required to make an annual standing payment of £S25 (the exchange rate is approximately £1 Sterling to £S1.8) to cover administrative and operational costs. Significantly, the choice of crops is left entirely to the individual farmer, as is the amount he wishes to sell (if any), to whom and where he wishes to sell it. Regulations requiring the cultivation of minimum areas of particular crops such as cotton or sugar do not apply. Farmers on the Khogalab irrigation scheme, therefore, have no direct constraints operating upon their choice of crops.

The survey of 85 farmers revealed eight different crops being cultivated (Table 9). In terms of numbers of farmers cultivating each, three distinct groups of crops emerge. In the first, bersim (Medicago sativa, or lucerne), vegetables and abosabayn (Sorghum vulgare) are, in eah case, grown by more than 62% of the surveyed farmers. Both bersim and abosabayn are fodder crops, and are grown with a minimum of input application. The second group comprises bamia (Hibiscus esculentis, or okra) and lubia (Dolichos lablab, or hyacinth-bean) with about 20% of farmers growing them, whilst the third group comprises three crops grown by only an insignificant minority of farmers. Few farmers, however, grew only one crop;

Table 9   Cropping Pattern

Crop	Farmers cultivating the crop		Citing as "most important" crop	
	No.	%	No.	%
Bersim	64	75.3	43	50.6
Vegetables	54	63.5	33	38.7
Abosabayn	53	62.4	5	5.9
Bamia	17	20.0	2	2.4
Lubia	15	17.7	1	1.2
Dura	3	3.5	1	1.2
Bananas	2	2.4	0	0.0
Melons	1	1.2	0	0.0
			85	100.0

N = 85

indeed, out of the 85 farmers in the survey, only 21 (24.7%) chose to grow only one crop on their farm, whilst 27 (31.8%) grew two crops, 20 (23.5%) grew three crops, and 17 (20.0%) grew four, five or six crops.

In the course of the survey, farmers were asked to identify their "most important" crop. To clarify the question in farmers' minds, it was suggested that this could be identified as the crop they would least like to lose, either because it would mean loss of income or loss of food for the family. The results in Table 9 show that two crops dominate, accounting for 76 out of the 85 responses, with bersim being cited as "most important" by half the farmers in the survey. When the six crops are aggregated into two groups, fodder crops and food crops, 48 farmers cited fodder crops as their "most important" (bersim and abosabayn), and 37 cited food crops (vegetables, bamia, lubia and dura).

There is a significant commercial orientation to agriculture in the study area with most farmers selling more than half their crop production. In the extreme cases of bersim and abosabayn, 93 per cent and 90% respectively of farmers sold all of their harvested crop. The massive urban demand of the Three Towns conurbation or Khartoum, Khartoum North and Omdurman only 35 km to the south along a tarmac road is clearly a crucial factor. This aspect of agriculture in the area is important in that farmers are faced with a choice between producing crops for consumption by the household and crops for sale in the market-place, conflicting demands which the farmer needs to reconcile as part of his decision-making process.

METHODOLOGY OF THE STUDY - REPERTORY GRID ANALYSIS

In an attempt to reduce interviewer interference and to gain insights into how farmers take decisions about their choice of crop, it was decided to use the technique of repertory grid analysis, although with modifications to suit the circumstances of the study. Repertory grid analysis is firmly rooted in personal construct theory (Kelly 1955) and it allows the respondent to build up a set of constructs relevant to his own intra-personal space (Bannister and Mair 1968; Fransella and Bannister 1977; Slater 1976; Shaw 1981). The technique has the attraction of allowing the respondent to identify his own constructs, this being achieved by presenting the respondent with a series of stimuli, or elements. Elements are conventionally presented in groups of three, or triads, the respondent being asked to judge which two are most similar and why they are different from the third element. This reason (or, indeed, reasons) constitutes a construct.

All the elements are presented to the respondent in every possible combination of triads, and a range of constructs is created. The elements and constructs are then arranged such that all the elements are arrayed along a horizontal axis, and the constructs along a vertical axis. The grid is then completed by grading every construct with reference to every element on a 5-point, 7-point or 10-point scale, the most usual being a 7-point scale with, for example, zero being irrelevant or unimportant, and seven being highly relevant or highly important, and the points in between representing gradations between these two polar extremes. Crucially, the technique allows the respondent to identify and grade those constructs which he sees as important and relevant, and not what the researcher thinks is important and relevant. This distinction is vital.

Repertory grid analysis, although designed at the outset by psychologists as a method of helping their patients to understand themselves better, has been used in the field of environmental perception quite successfully, for example in the perception of countryside locations in north-west England and North Wales (Palmer 1978), the perception of shopping outlets (Hudson 1974), and the presentation of different sets of proposals for shopping centre redevelopment in London (Stringer 1976). The technique has also been used in rural Third World situations, primarily in the field of perception. Townsend (1976,1977) used the technique successfully in her study of rainforest farmers and colonisers in Colombia, and Floyd (1977) also used the technique in his study of small-farmers' perceptions and motivations in agriculture in Trinidad. Within Africa, Barker et al (1977) and Barker (1979) employed repertory grid analysis in the identification of agricultural problems and the understanding and measurement of environmental images. To the present author's knowledge, repertory grid analysis has not been used in the study of farmers' choice of crops, apart from the notable exception of a study of agricultural decision-making in Warwickshire (Ilbery and Hornby 1983). A similar, but distinctively separate, technique of semantic differentiation has been used, but this lacks the subtlety of repertory grid analysis (for a fuller discussion, see Briggs, forthcoming).

With reference to the present study it was decided to present as elements those crops grown by the respondent, as well as any other crops grown by other farmers on the irrigation scheme, but not grown by the respondent himself. By asking him why he grew some crops rather than others, and why he considered one particular crop to be his "most important" crop rather than any of the others he grew, decision factors were elicited from the respondent as constructs. The onus was placed very much on the respondent to provide constructs, or decision factors, and, as a result, the number of constructs elicited by individuals in the pilot survey of 30 farmers varied between two and eleven.

A total of 18 different constructs, or decision factors was elicited by the pilot survey. These 18 factors were then placed in a standardised grid for the second part of the survey which was administered to a total of 85 farmers from the irrigation scheme, included within which were the 30 farmers of the pilot survey. Each farmer was asked to grade each of the 18 constructs on a 5-point scale with reference only to their "most important" crop (see Table 9), this limitation being imposed to reduce confusion on the part of the responding farmer. Zero was interpreted to be "very unimportant" and a score of 4 to be "very important", with scores of 1, 2 and 3 representing gradations of importance between the two poles. In addition, the 85 respondents were given the opportunity to identify further decision factors not included in the standardised grid, but none was forthcoming, suggesting that the 18 factors elicited in the pilot survey cover the range.

A list of decision factors and the aggregate scores for each are presented in Table 10. The aggregate score for each factor is calculated by summing all the scores given by each farmer on each of the factors. As the maximum score on any factor is 4, it follows that with 85 respondents the maximum possible score is 340 points (85 multiplied by 4); consequently, the top factor totalled 286 points, or 84.1% of the maximum score available.

Inspection of Table 10 reveals a range of decision factors, included within which are purely economic factors such as "the crop commands a good price", overtly social factors (for example, "the crop leaves me a lot of free time"), inertia factors ("the crop has always been grown here"), personal contact factors ("a friend/relative gave me the idea to grow the crop"), and factors which do not neatly fit into a particular slot; for example, "a lot of experience with the crop" could be interpreted as being a inertia factor or an economic factor, depending on particular emphasis. However, what is clear is the fact that by using repertory grid analysis a wide-ranging group of decision factors have been elicited which, in themselves, point to the complexity of the decision-making process in the choice of crops. Although a discussion of the interpretation of the factors presented in Table 10 is outside the scope of this paper, three general observations can be made. First, even in a relatively commercial-orientated area, non-economic factors such as experience, reliable yields and long history of cultivaton of the particular crop dominate farmers' decision factors. Second, the impact of personal communication would appear to be rather limited. Third, economic factors emerge as important decision factors only after secure cultivation of crops has been achieved. For a detailed discussion, see Briggs (forthcoming).

METHODOLOGICAL PROBLEMS AND ASSUMPTIONS

The main theoretical and methodological problems associated with repertory grid analysis and personal construct theory are thoroughly discussed in the literature (see, for example, Kelly 1955; Bannister and Mair 1968; Shaw 1981). However, a brief outline of the key problems experienced in this survey is of value.

The construction, form and content of the standardised grid for the main survey of 85 farmers provided both theoretical and methodological problems. As repertory grid analysis allows an individual to identify his own constructs, the way in which he sees his world, and as these constructs are personal and unique to that individual, clearly when a series of grids among a given population are administered, there are likely to be as many different non-comparable grids as there are members of that population; in the case of the present survey, that would be 85. Further, not only might the type of construct elicited be different, but the absolute number elicited is also likely to differ. In the pilot survey of the present study, the number of elicited constructs from the 30 farmers varied between two and eleven. Clearly, under such circumstances, comparability of grids is impossible and the standardised grid becomes a necessity. This, however, entails a number of problems. Firstly, the construction of the standardised grid requires selection on the part of the

Table 10 Aggregate scores for decision factors

Decision factor	Score	Score as percentage of maximum
1. A lot of experience with the crop	286	84.1
2. The crop has always been grown here	283	83.2
3. The crop generally gives a good yield	269	79.1
4. There is a reliable market for any of the crop sold	255	75.0
5. The crop gives a good income relative to effort	253	74.4
6. It is an easy crop to grow	247	72.7
7. The crop leaves me a lot of free time	238	70.0
8. The crop commands a good price	235	69.1
9. The crop does not need much hired labour	218	64.1
10. The growing of the crop does not risk my livelihood	210	61.8
11. The crop does not need much money to grow it	187	55.0
12. Before planting, I have a good idea of the market price	183	53.8
13. A friend/relative gave me the idea to grow the crop	176	51.8
14. The crop is not easily damaged by birds/animals	176	51.8
15. I grow the crop because my friends grow it	128	37.7
16. My family like the taste of the crop	108	31.8
17. I have been trained to grow the crop	101	29.7
18. An extension officer persuaded me to grow the crop	80	23.5

Total number of respondents - 85

interviewer as to which constructs to include and which to reject, thus introducing a degree of interviewer interference, something which repertory grid analysis is designed to reduce. The dilemma, therefore, is one of having to compromise between, on the one hand, freely-elicited individual grids which are non-comparable but minimise interviewer interference, and, on the other, standardised grids which do allow comparability but at the same time increase interviewer interference. Despite misgivings, however, Fransella and Bannister (1977) do not regard the selection of constructs as an important problem for the interviewer as long as the constructs presented in the standardised grid are elicited from individuals within the population:

There is no definite evidence to indicate that you should not provide constructs for a grid. On the contrary, there is some evidence to suggest that results using provided constructs produce meaningful results ... and are significantly related to individuals' behaviour.
(Fransella and Bannister, 1977, p. 107).

Easterby-Smith (1981) shares this view with the added proviso that the respondent has an adequate understanding of what the constructs mean.

In the present study, all 18 constructs elicited in the pilot survey were included within the standardised grid, on the grounds that they were elicited from within the population, and that the 30 farmers themselves from the pilot survey were to be included within the full survey of 85 farmers for the administration of the standardised grid. None of the constructs elicited in the pilot survey were excluded from the standardised grid, but it should be pointed out that there was no major overlap between constructs, a fortunate state of affairs which is unlikely to be consistently repeated in future surveys. Additionally, it should be repeated that farmers in the main survey were given the opportunity to add to the standardised list of constructs any constructs personal to themselves.

The problem of including all the constructs emerged at the end of the main survey. Even if only one of the pilot survey respondents had produced a construct, it was included in the standardised grid. Clearly, the construct was important in that individual's choice of crop, but it may not have been so among the vast majority of farmers. This appears to be the case for personal contact factors such as "an extension officer persuaded me to grow the crop", which showed up very badly in the standardised grid (Table 10). On further reflection, this may not be altogether surprising as this type of decision factor is more likely to be at its most relevant when farmers are selecting new crops. Further inspection of Table 10 shows that inertia factors come out strongly, very much at the other extreme to innovation. Therefore, by including all elicited constructs, it is likely that some degree of interrelationship between factors will occur, to the extent that "a lot of experience with the crop" and "an extension officer persuaded me to grow the crop" can be interpreted as two sides of the same coin.

A further problem associated with the standardised grid concerns the selected constructs themselves. A basic assumption of repertory grid analysis is that a grid represents an individual's system of personal constructs. This assumption, however, has to be relaxed in a standardised grid when a respondent is presented with a series of constructs, some of which, or, indeed, many of which, may not have been elicited by himself in a personal, non-standardised grid. This raises important questions and, in particular, the extent to which constructs are personal and the extent to which they reflect basic common beliefs and perceptions of the larger group. In the context of standardised grids, support for the latter question needs to be strong. The problem, however, may not be quite as severe as it first appears, because if the respondent in the standardised grid is of the opinion that a particular construct is irrelevant or that he would not have identified it himself, then he has the option of rejecting the construct completely by grading it with a score of zero. Furthermore, the use of a standardised grid may in fact give a better indication of decision factors, particularly where the number of constructs elicited freely from an individual is more a function of articulacy than anything else; thus the less articulate individual benefits from having constructs elicited from more articulate fellow-farmers presented to him for grading. This naturally assumes that a particular construct means the same thing to different people (Bannister and Mair, 1968).

The identification of the decision-maker (and therefore the respondent to the standardised grid) can present difficulties. In the present study, the head of household was almost invariably the decision-maker, at least in the sense that the final decision was taken at this level. This, however, need not always be so. The head of household may have a non-agricultural job elsewhere and the day-to-day decisions, and even medium- and long-term decisions, may be made by some other member of the household. Equally, the decision as to what crop to grow may be taken by several household members together, thus raising the question that a household's collective decision may not be identical with one made by the head himself. Care, therefore, must be exercised to ensure that the standardised grid is administered to the most relevant person.

In the operation of the standardised grid, two practical issues had to be confronted. Firstly, it was decided to use grading rather than ranking in the evaluation of constructs. Ranking of large numbers of constructs becomes difficult, particularly in the middle range when a construct is not obviously good or bad, relevant or irrelevant, or important or unimportant to the individual. Grading, on the other hand gives the respondent greater flexibility and, in practice, effectively allows constructs to be tied in importance, arguably more in accordance with reality than attempting to rank each between first and eighteenth. Secondly, respondents were asked to grade decision factors in the standardised grid in relation to their most important crop only. It was not possible to grade factors for each crop grown because the biggest mix of crops on any one farm in the survey was six, whilst 21 farmers cultivated only one crop. To have asked respondents to grade each construct in terms of each crop that they grew would have re-introduced the problem of non-comparability of grids. By using only one element ("the most important" crop), however, the use of the Grid Analysis Package (Slater 1972, Chetwynd 1974) a computer package designed specifically for this type of data, was severely curtailed; indeed, it has been pointed out that with fewer than six elements in a grid, the results produced by using the Grid Analysis Package can become distorted (Easterby-Smith 1981). This was regrettable, but necessary, if meaningful results were to be produced from the farmers themselves.

CONCLUSIONS

From the above discussion, three general points can be made. Firstly, repertory grid analysis is successful in reducing interviewer interference and, by transferring the initiative to the respondent, it helps to overcome some of the difficulties experienced by the interviewer, particulrly in cross-cultural situations. In addition, by allowing the respondent to talk round constructs, it can provide a useful insight into those processes involved in the choice of crops, but still within a relatively well-defined structure. Secondly, on the negative side, the technique is time-consuming to administer (although far less so once the standardised grid has been established), and can be difficult to train field assistants to use. On the positive side, though, respondents invariably found it an interesting, and in some cases rewarding, exercise in which to participate, and on more than one occasion, a respondent finished off the interview session by saying that he now understood far more about his own farming activities. Finally, there is still a marked need for many more studies to be undertaken using this technique, to explore its limitations and weaknesses, to test its validity and to evaluate its findings, and particularly so in the area of Third World agricultural studies. Repertory grid analysis would appear to be a useful technique in the understanding of Third World agricultural activity, but there is still some way to go before this can be confirmed.

ACKNOWLEDGEMENTS

I would like to thank the Nuffield Foundation, the Carnegie Trust for the Universities of Scotland, the University of Glasgow Research and Travel Support Committee and the Dudley Stamp Memorial Trust for providing research grants for fieldwork in Sudan. In addition, I would like to thank the Department of Geography, University of Khartoum for providing assistance with the project.

REFERENCES

Bannister D. and Mair J.M.M. (1968) The evaluation of personal constructs. London.

Barker D. (1979) Appropriate methodology: an example using a traditional African board game to measure farmers' attitudes and environmental images. IDS Bulletin, 10, 37-40.

Briggs J. (forthcoming) Farmers' choice of crops: a study of irrigating farmers in central Sudan. Occasional Paper Series, Department of Geography, University of Glasgow (publication date mid-1984).

Chambers R. (1983) Rural development: putting the last first. London.

Chapman G.P. (1974) Perception and regulation: a case study of farmers in Bihar. Transactions Institute of British Geographers, 62, 71-93.

Chetwynd J. (1974) Generalised grid techniques. St. George's Hospital, London.

Easterby-Smith M. (1981) The design, analysis and interpretation of repertory grids. In Shaw M.L.G. (Ed.) Recent advances in personal construct technology, 9-30.

Floyd B.N. (1977) Small-scale agriculture in Trinidad: a Caribbean case study in the problems of transforming rural societies in the tropics. University of Durham Department of Geography Occasional Paper (New Series), No. 10.

Fransella F. and Bannister D. (1977) A manual for repertory grid techniques. London.

Hudson R. (1974) Images of the retailing environment. An example of the use of repertory grid methodology. Environment and Behaviour, 6, 470-494.

Ilbery B.W. and Hornby R. (1983) Repertory grids and agricultural decision-making: a Mid-Warwickshire case-study. Geografiska Annaler, 658, 77-84.

Kelly G. (1955) The psychology of personal constructs. London.

Lele U. (1975) The design of rural development: lessons from Africa. Washington D.C.

Long N. (1977) An introduction to the sociology of rural development. London.

Palmer C.J. (1978) Understanding unbiased dimensions: the use of the repertory grid methodology. Environment and Planning A, 10, 1137-1150.

Shaw M.L.G. (Ed.) (1981) Recent advances in personal construct technology. London.

Slater P. (1972) Notes on INGRID '72. St. George's Hospital, London.

Slater P. (Ed.) (1976) Explorations of intra-personal space. Chichester.

Stohr W.B. and Taylor D.R.F. (Eds.) (1981) Development from above or below? The dialectics of regional planning in developing countries. Chichester.

Stringer P. (1976) Repertory grids in the study of environmental perception. In Slater P. (Ed.) Explorations of intra-personal space, 183-208.

Townsend J.G. (1976) Farm "failures": the application of personal constructs in the tropical rainforest. Area, 8, 219-222.

Townsend J.G. (1977) Perceived worlds of the colonists of tropical rainforest, Colombia. Transactions Institute of British Geographers, 2, 430-458.

World Bank (1981) Accelerated development in sub-Saharan Africa: an agenda for action. Washington D.C.

IRRIGATION AS HAZARD: FARMERS' RESPONSES TO THE INTRODUCTION
OF IRRIGATION IN SOKOTO, NIGERIA

W.M. Adams

INTRODUCTION

Interest in the possibilities of irrigation in the Sokoto Valley in northwest Nigeria dates back to 1917 and 1918 when several small water conservation schemes were built within the floodplain above Sokoto city. These fell into disuse, but the interest persisted, with the visit of an irrigation engineer in 1923, and the construction of a pilot project at Kware a few years later. Despite many problems, activities continued at Kware (see Palmer Jones 1981), and eventually the river basin became the subject of an extensive study by the Food and Agriculture Organisation in the 1960s (FAO 1969). The main stimulus for seeking this study was the government's fear of soil erosion, desiccation and, less fully worked-out, a long-standing concern for 'the advance of the desert' in the north of Nigeria (Adams 1983).

In a national report in 1966 the FAO had strongly argued the central importance of irrigation to the Nigerian economy (FAO 1960), and the Sokoto study endorsed this by recommending a series of dams and irrigation schemes to control the strongly seasonal flows of the basin's rivers and make new and more productive use of the water resource schemes.

The first project at Bakolori on the River Sokoto was begun in the early 1970s. The Bakolori Project as designed was different from that originally planned by the FAO, being increased in size by 2.5 times, but its aim was ostensibly unchanged: to control flood flows in the river and use the water for irrigation to offset the arid and unpredictable environment of the Sokoto basin. Ironically, in two ways the scheme has increased the uncertainty and, temporarily at least, reduced the productivity of the valley. First, there has been a significant loss of productivity in the flood plain downstream of the dam and irrigation scheme (Adams 1983). Second, the introduction of irrigation in the scheme itself has brought uncertainty and loss of production to farmers. This impact of irrigation is directly comparable to the more familiar problem of drought in its effect on farmers, and has led to the adoption of familiar strategies developed to survive the drought hazard.

THE DROUGHT HAZARD

The dry Sudan Zone of northern Nigeria is a relatively hazardous environment for the traditional cultivator. It has low rainfall, a single short rainy season and a history of periodic drought. The Sokoto Valley has a wet season of about 4 months (June-September) and an annual rainfall of between 700 and 900 mm. However, annual rainfall is variable, and major droughts have occurred a number of times since the turn of the century, notably in 1913 (Grove 1973) and in the 'Sahel Drought' of 1972-74 (Mortimore 1973, Apeldoorn 1978). There have been widespread famines and conditions of food scarcity in a dozen years since 1900, usually associated with drought (Apeldoorn 1981).

Hausa farmers have developed a series of strategies for dealing with the hazards presented by the vagaries of the natural environment. In dry land agriculture, intercropping is used as a risk-avoidance strategy (Abalu 1977), and crops such as bulrush millet and sorghum are adapted to dryland conditions: sorghum, for example, has waxy leaves which curl under moisture stress, and a low evapotranspiration rate (Cobley and Steel 1970). In floodplains like the Sokoto, rice and sorghum varieties adapted to different soil and flooding conditions are grown to ensure some production whatever conditions occur. (Adams in prep).

Farmers also engage in a number of off-farm activities for cash incomes which provide a cushion in times of drought. Apart from craft activities, in the Sokoto Valley a number are traders, the term covering a range of activities from long-distance itinerant trading to local peddling of petty goods. For others, a major source of cash income is dry season labour circulation, cin rani. Men go as far afield as Ghana and the towns of southern Nigeria from the Sokoto Valley (Abdu 1980, Adams 1983).

In drought years the agricultural adaptations to the aridity and unpredictability of the environment are no longer sufficient to cushion farmers from the conditions. Even the productive floodplain farming will fail with reduced rainfall. In such years the off-farm activities provide a framework for survival, as farmers have to look to wage-employment away from the village to survive. Normal levels of cin rani increase as more farmers move away in search of work, the familiar locations attracting many who normally do not go. Studies in Hausaland in the 1972-74 drought show how the existing patterns of circulation in villages become in drought years the pattern for movements by greatly increased numbers of people, both for one season and for longer periods (Mortimore 1979).

The traditional pattern of response to the hazards of drought seems to be one of adaptation within agriculture, but when this fails a move out of agriculture, and indeed out of Hausaland, in search of cash incomes in other activities. The introduction of irrigation in the Sokoto Valley has, temporarily at least, elicited a similar set of responses.

IRRIGATION IN THE SOKOTO VALLEY

Work on the Bakolori Agricultural Project began in the early 1970s (Impresit 1974), and construction of the dam began in 1974. The project was officially inaugurated in 1978, although work on the irrigation area continued for some years after that. The project involves a 414 Mm dam on the Sokoto River, a 10 Km long concrete-lined supply canal to carry water to irrigate 27,000 ha of terrace and floodplain land near the town of Talata Mafara. A pipe-bridge carries water over the river to irrigate land on the right bank, and a buried pipeline and pumpstation at the dam carries water up to the 2,000 ha Jankarawa Scheme above the North shore of the reservoir close to the town of New Maradun where many of the evacuees from the reservoir are settled. Irrigation is partly by surface delivery, for which extensive land levelling has been required, and partly by sprinkler using movable pipes and buried hydrants. The irrigation area was more than 80% cultivated before the development began

The Bakolori Project is still not fully commissioned. Construction work was about 70% complete by 1981, but irrigated cropping lagged far behind. Preliminary studies have suggested that there are both social and economic problems within the project area itself. Oculi (1980), Wallace (1979) and Igozurike and Diatchavbe (1982) highlight some of the social problems associated with land policy on the scheme, and Etuk and Abalu (1982) the shortfalls which have occurred in acreage and producion targets. The problems are similar to those being encountered in other large irrigation schemes in the north of Nigeria such as the South Chad Irrigation Project and the Kano River Project (Carter 1981, Wallace 1981), and are similar also to those experienced with irrigation since its first inception in Nigeria (e.g. Palmer Jones 1981).

THE HAZARDS OF IRRIGATION: LAND

Studies were carried out in 1980 on one village on the Bakolori Project called Birnin Tudu. The work concerned primarily the relation between cropping patterns and natural flooding in the floodplain, but in the course of interviews and informal conversations a considerable amount of information was given about the irrigation scheme itself. The farmers in Birnin Tudu had two main problems in 1980. The first related to the expropriation of their farmland and its reallocation. Some of the problems of land tenure on northern Nigerian schemes are discussed by Alan Bird (this volume). At Bakolori in both surface and sprinkler areas land was expropriated for engineering work before, in theory, being handed back for irrigated cultivation. Before development began more than 80% of the project area was already in use for rainfed cultivation.

The problem in Birnin Tudu was the length of time that farmland remained unavailable for cultivation during development for irrigation. Expropriation began in 1977 on Birnin Tudu land, and reallocation had still not been completed by December 1980. Designs for Intake E. Right, where most of the village's land lay, were approved in April 1977. The Contractor hoped to start work on land development in that year, and was dismayed to find crops planted: after discussion, he restricted himself to work on the canal lines, plant yard and project offices. In the 1978 wet season however, all cropping was banned in advance, and land clearance began in May of that year. Despite the ban, crops were planted again, and in June farmers, angry at the destruction of growing millet, stopped the Contractor from working.

The farmers planted because they needed to: there was no alternative source of food or income, and compensation was not payable for crops which were not planted. Also, it seems that they did not believe the cropping restrictions meant anything, because the previous year the Contractor had failed to work in some areas in nearby Intake

C, even though a cropping ban had been placed on the area to allow him to do so. Farmers who had obeyed the cropping ban were angered that they had missed out on the chance to raise a crop, and others were unwilling to be caught in the same way.

The problems in Intake E continued into 1979. Table 11 shows the proportion of land on which cropping was banned, and that over which cropping was actually prevented, in 1978 and 1979: cropping was impossible over 60% of the area in 1978 and 40% the next year. Land levelling began in the area in July 1979, and reallocation on paper, although not on the ground, began in October of that year. Actual reallocation of land proceeded much more slowly, falling steadily behind the engineering work of land preparation. Furthermore, the land taken in 1977 for the earliest stages of construction would not be handed back until the whole intake was reallocated as a unit. By December 1980 land had still not been formally handed back to the farmers of Birnin Tudu so that they could cultivate. Their fields had lain barren for up to 3 years at that stage.

These problems of reallocation were compounded by disputes over the payment of compensation for economic trees and improvements to land, both non-payment and under-payment being alleged by farmers. Arguments over compensation on the project as a whole began in 1978 (Muazu 1979), but became steadily more acrimonious in 1978 and 1980 with a prolonged blockade of the project culminating in a police action on 26 April 1980 involving between 25 and 120 deaths (Concord 1980, Adeyanju 1980). The official death toll was 25 dead, 23 of them farmers. This did not of course solve the problem of compensation: in November 1980 an official announcement by the SRBDA stated that the compensation process was expected to last several more years (Salihu 1980).

Birnin Tudu was the only substantial settlement assaulted by the police, in April 1980, and apart from an undetermined number of arrests (including that of the village head) and deaths, the resulting fire destroyed houses and grain stores as well as consumer goods (bicycles, motorcycles, sewing machines) and business properties, particularly the small shops and tailors' kiosks on the accessible road side of the village. At the time of the survey, therefore, a number of the items purchased with the compensation money had been destroyed, and considerable grievance was felt against the project for this reason.

UNPREDICTABLE FLOODING

The second problem faced by the farmers in Birnin Tudu in 1980 concerned the flooding of the small piece of floodplain or fadama land left to them near the village. This land, like the rest of the floodplain, had been subject to annual inundation by floodwaters before the construction of the Bakolori Dam. This allowed a productive flood-recession agriculture to be practised, based on rice in the wet season, and vegetable crops in the dry season. Before the dam was built, 92% of the area flooded. Dam construction reduced flooding to 33% of the area, with corresponding uncertainty for the farms, change from rice to less productive soya and millet cropping patterns, and loss of dry season cultivation (Table 12).

Table 11 Prohibition and Loss of Crops on Intakes E and D

	1978	1979
% Area where cropping prohibited	65%	22%
% Area over which crop lost	60%	42%

(Total area intakes 2,000 ha)

Table 12 Flooding of the Birnin Tudu Fadama

	Area (total 146 km)	
	% flooded by area	% flooded over 1 month
Before Dam	92%	19%
After Dam (1979)	33%	2%
1980	74%	44%

125

The fadama area was excluded from the irrigation scheme because of drainage problems, but the terrace land above it was developed, levelled and drained. Unfortunately, the drains debouched on to the fadama at two points, and because of the area of their catchment and the engineered efficiency of the drainage, discharge in the drains after a storm was rapid and of considerable magnitude. Extensive flooding of the fadama therefore occurred once again in the wet season of 1980 (Table 12): 74% of the fadama was flooded. The floodwaters from the terrace land covered growing dryland crops unable to withstand inundation, depositing large amounts of sandy sediment. Furthermore, the water lay in the deeper pools and hollows for far longer than the old river flooding: it lay for more than a month in 44% of the area, compared with about half that before the dam. Two months after the first inundation 17% of the area was said to be still flooded to such a depth that a man might not cross.

The farmers largely gave up trying to crop the fadama area in the 1980 wet season, and a wide area lay untended or carried crops destroyed by the flood. Of course the effect of the riot in April and its aftermath are likely to have contributed to the area left uncropped, since for a number of weeks in April the villagers fled the area, but the farmers themselves claimed that the flooding was deliberate and part of the Project's plans, and said they saw no reason to plant crops only to have the land taken over in some way. Thus in the wet season of 1980 few farmers harvested crops from the Birn Tudu fadama, even though in theory it was available to them. It should be noted that plans were drawn up by the engineer at Bakolori to prevent the recurrence of such flooding in subsequent years, although the success of implementation is less clear.

RESPONSES TO THE HAZARDS OF IRRIGATION

To the farmers in Birnin Tudu the initial years of irrigation development were hazardous. They were prevented from farming in the area being developed for irrigation for several years, and in 1980 lost crops in the small area still left to them. Interviewees in Birnin Tudu were asked open-ended questions about their attitude to the project. 75% of respondents in 1980 said they had, as yet, received no benefit from Bakolori: 50% cited specific grievances, among which the apparent loss of their land features prominently (44 respondents, November 1980 survey). In part the farmers responded to this situation by using the familiar strategies of times of drought. First, within agriculture, they became opportunistic, and it was possible to see rice planted in flooded drainage ditches on tudu (upland) soils in areas worked over by the Contractor but not reallocated. In other cases small bunds were built to retain water on such land. Second, the farmers turned away from cultivation and went elsewhere in search of work. Levels of cin rani (seasonal migration) in Birnin Tudu were high, with 53% of households containing migrants. In two other villages on the Bakolori Project the incidence was 65% and 41% of households, in contrast to figures from villages in other parts of the Sokoto floodplain of 29% (Burbawa, near Sokoto) and 13%

(Rane, a smaller and more remote village).

The third response of the farmers was to move out of farming more permanently. In the sample of 44 men interviewed, 11 (25%) said they had benefitted from the scheme at Bakolori. Of these 8 (18%) attributed their relative success to employment with agencies involved in the development. One man himself had a job as a messenger with one of the firms, 5 had close relatives in their house with wage-paid jobs, all unskilled. A further two men had held jobs with the project in the past and had used the money to set themselves up in business, one as a tailor, one as a cloth trader. Even the men who had no direct contact with the project benefitted indirectly - one was Sarkin Pawa, the chief butcher, one a trader in grain and cloth who frequently worked for the village head, and one a cloth trader. All benefitted from the better road access to the village, the greater amount of cash circulating from compensation payments and the opportunities for trading presented by the project.

There was no doubt among interviewees that the best prospects were associated with wage employment. One man, revisited in December 1981 by which time irrigation water was available on some at least of the village's land, agreed that farming was not possible again, but stressed that the real problem for him was that none of his family had a job with the project: without that he felt his prospects were poor. The importance of contacts with the project can perhaps be best shown by presenting case studies of two men whose relative success in the village can be attributed to that source.

The first man, Mu'azu, was Sarkin Tasha, the man in charge of motor transport to and from Birnin Tudu. There was no formal motor park in the village, but he worked at the roadside arranging journeys and fares. Before the project and the new road, vehicles came to Birnin Tudu only once a week to carry people to Talata Mafara market. In 1980 a great many vehicles plied the road to the village and the project or passed the village to cross the river by the new bridge to the settlements of the north. Mu'azu derived considerable income from this position of Sarkin Tasha, but also made contacts with people on the project (as he put it "I know a lot of big men now") and had become something of a broker between the village and the project, with opportunities for sundry business transactions. He had lost possessions in the 1980 fire, but was sanguine about his prospects because of the opportunities offered by these contacts. The second man, Halilu Mashal, had a shop in Birnin Tudu selling general trade goods bought as far away as Kano. This was burned in the fire in 1980, but already by November of that year he had established a small kiosk at the entrance to the village where he sold sugar cane sticks. He had been a petty trader for some years, selling small items such as sweets acquired in Sokoto on cin rani, when a job on the project enabled him to expand. He first became a labourer, then a plasterer at the project offices and finally a night guard at the residential complex near Talata Mafara. With the money he earned he opened a substantial shop, and had plans to do so again as soon as he could.

Clearly in both these instances men had done well by taking advantage of the opportunities offered by the development of the project. Equally clearly, the only people in the village who were doing well owed their success to the same source. The situation may change as irrigated farming expands, but at the interim stage when the survey was carried out it was those men who opted out of farming and into trade or wage employment who were doing well. Clearly, given a sufficient disincentive to farm (loss of production or immobilisation of land) and a range of alternative opportunities, some men at least will adapt themselves to new livelihoods. Further study is needed on those who do not or cannot adapt in this way.

IMPLICATIONS

At Bakolori the introduction of irrigation has obviously been traumatic and hazardous for the farmers involved. The echoes of the riots and the deaths are still sounding, indeed growing with the telling (cf. Soyinka 1984). It is ironic, and potentially disastrous for the scheme, that one of the farmers' chief responses to their troubles has been to give up farming and seek incomes elsewhere.

Of course, the problems at Bakolori could be looked on simply as teething troubles, and this accounts for the result of "design period bias" in research (Chambers 1981), unhelpfully focussing on the most difficult period in the life of an irrigation scheme. It is true these observations refer to the initial period of introduction of irrigation only, but this initial "launch" period is of critical importance in determining the farmers' expectations of the scheme and willingness to be involved. At Bakolori the opportunity to benefit from this period was lost. Although it is likely that as the bottleneck of land reallocation at Bakolori eases and irrigation farming becomes possible, farmers will become more involved, it is certain that their attitudes will remain soured by their initial experiences. Even if irrigation at Bakolori were to prove of clear benefit to the farmers (and this remains to be demonstrated) a considerable effort in extension would be required to convince them of this fact.

Bakolori was built rapidly and, judged solely as a construction problem, efficiently. Unfortunately the socio-economic planning frameworks necessary to make the project practicable were not so well worked out, and the consequent problems have proved singularly intractable to deal with. It seems a common problem with irrigation schemes that the "hardware" of engineering is far better developed than the "software" of socio-economic planning. At Bakolori at least the latter has proved of central importance. One might conclude that a larger socio-economic input on the project, and perhaps better methodological inputs by social scientists in general would enable schemes like Bakolori to be made to work smoothly. This is the common solution, that doing more socio-economic studies wil somehow make the scheme a success: thus better "software" will make the "hardware" work. In fact the experience at Bakolori suggests something rather different. If schemes are going to work at all they must start with the people involved and work back from their needs and aspirations, not try to fit people into a fixed and predetermined plan and engineering schedule. This means that the form of the engineering aspects of the development must be open to definition by the social and economic perspectives of the local people, and not the other way round. This sounds easy, but it runs counter to all established procedures and assumptions of project conception, design, funding and appraisal. It promises to be remarkably difficult to achieve. If it is not attempted, we cannot be surprised if schemes sometimes fail spectacularly to achieve success.

REFERENCES

Abalu G.O.I. 1977 A note on crop mixtures under indigenous conditions in Northern Nigeria. Samaru Research Bulletin 276, Zaria, Nigeria.

Adams W.M. 1983 Downstream impact of river control, Sokoto valley, Nigeria. Unpublished Ph.D. University of Cambridge.

Adams W.M. In press. Floodplain cultivation and traditional water control in the Sokoto valley, Nigeria. Geographical Journal.

Abdu P.S. 1980 Changes taking place in spatial patterns of mobility: the Sokoto example. Paper to seminar Change in Rural Hausaland, Bagauda Lake Hotel, Nigeria. February 29th to March 1st, 1980.

Adeanju B. 1980 The Bakolori affair: what really happened. Daily Times (Lagos) 10 June 1980.

Apeldoorn G.J. van (Ed.) 1978 The aftermath of the 1972-74 drought in Nigeria: proceedings of a conference held at Bagauda, April 1977. Centre of Social and Economic Research, Ahmadu Bello University, Zaria.

Apeldoorn G.J. van 1981 Perspectives on drought and famine in Nigeria. George Allen and Unwin, London.

Carter R. 1981 Learning from irrigation experience in Northern Nigeria. Paper to International Commission on Irrigation and Drainage Workshop 4-6 April 1981, Southampton, UK.

Chambers R. 1981 In search of a water revolution: questions for canal irrigation management in the 1980s. Water Supply and Management 5(1): 5-18.

Cobley L.S. and Stede W.M. 1976 An introduction to the botany of tropical crops (2nd Edition) Longman, London.

Concord 1980 Needless deaths. National Concord (Lagos) 8th July 1980.

Etuk E.G. and Abalu G.O.I. 1982 River basin development in Northern Nigeria: a case study of the Bakolori Project. Proc. 4th Afro-Asian Regional Conference of the International Commission on Irrigation and

Drainage, Lagos, Nigeria. Volume II (Publ. 26): 335-346.

FAO 1966 Agricultural development in Nigeria 1965-1980. Food and Agriculture Organisation, Rome.

FAO 1969 Soil and water resources survey of the Sokoto valley, Nigeria: final report. Food and Agriculture Organisaton, Rome.

Grove A.T. 1973 A note on the remarkably low rainfall of the Sudan Zone in 1913. Savanna 2(2): 133-138.

Igozurike M. and Diatchavbe O. 1982 The social cost of irrigation: the case of Bakolori. Proc. 4th Afro-Asian Regional Conference of the International Commission on Irrigation and Drainage, Lagos, Nigeria. Volume II (Publ. 27): 347-352.

Impresit 1979 Bakolori Project: first phase of the Sokoto-Rima basin development. Final Report. Impresit and Nuovo Castoro, Milan and Rome.

Mortimore M. 1973 Famine in Hausaland 1973. Savanna 2(2): 233-234.

Mortimore M. 1979 Natural and social factors in the problems of the Sahel, with particular reference to Northern Kano state, Nigeria. Nordic. Assoc. for Development Geography, Copenhagen.

Muazu A. 1978 Villagers stage violent demonstration - one killed, 17 injured. New Nigerian. 7 September 1978.

Oculi O. 1980 The political economy of the planning of the Bakolori Irrigation Project 1979-1980. Paper to Centre of Social and Economic Research. Ahmadu Bello University at Yankari, 12-16 May 1980.

Palmer-Jones R.W. 1981 How not to learn from pilot irrigation projects: the Nigerian experience. Water Supply and Management 5(1): 81-105.

Salihu I.N. 1980 Bakolori Dam Project - compensation to last until 1983. New Nigerian 1 July 1980.

Soyinka W. 1984 Why Shagari couldn't last. The Guardian 3 January 1984.

Wallace T. 1979 Agricultural projects and land in Northern Nigeria. Review of African Political Economy 17:59-70.

Wallace T. 1981 The Kano River Project Nigeria: the impact of an irrigation scheme on productivity and welfare. pp. 281-305 in Heyer F. Roberts P. and Williams G. (Eds.) Rural Development in Tropical Africa. MacMillan, London.

SIMULATION AND MODELLING FOR TRAINING IN IRRIGATION MANAGEMENT

M.A. Burton and T.R. Franks

INTRODUCTION

It is widely recognised that many irrigation schemes in developing countries have not performed as well as expected. In older schemes the procedures for management have remained, though the ability to control them has deteriorated: productivity in newer schemes is constrained by a lack of trained staff. Throughout the spectrum of irrigated agriculture there is no doubt that significant increases in productivity can be obtained by improving management. This applies particularly in Africa where there is often a shortage of skilled manpower and little experience of irrigation.

The basic objective of irrigation management is to ensure that schemes operate effectively, so that the best use is made of the available resources. As the rate of new development decreases, increasing interest is focussed on training for improved irrigation management. Simulation and models have been found to be useful as components in a training programme, their use in training in irrigation management is expected to increase.

The distinction drawn between models and simulations in this paper should be noted. Simulations are intended as likenesses, rather than models, of the real life situation. Thus in a training situation it might be acceptable to exaggerate certain features of a real life situation in order to make a point. Models on the other hand should reflect as closely as possible the actual real life situation and as a consequence are usually more sophisticated than simulations.

THE SCOPE OF IRRIGATION MANAGEMENT TRAINING

The requirements for irrigation management training can be divided into two parts. The first is training in the scientific knowledge and physical techniques which must always form the basis of efficient operation and management. This encompasses aspects from an understanding of soil and management to the characteristics of gate operation, and includes such matters as monitoring, budgeting and cost control. The second part is the development of new managerial skills such as critical planning, interpretation of available data, organisation, staff motivation and a better understanding of human relations and needs. The training approach within these two parts may be further sub-divided into categories of staff who analyse data and make decisions and a separate group who collect data and implement instructions.

The scope of training thus includes at least the following major areas:-

Objectives of irrigation development;
System design and operation;
Soil - plant - water relationships;
Field irrigation;
Monitoring and evaluation;

Financial management and planning;
Motivation and communication.

Such training may be required for personnel with a wide range of education and experience. They may extend from senior scheme management through to canal superintendents, agricultural extension officers, gate and pump operators, and water bailiffs.

The discussion in this paper is restricted to the training of management staff. As a consequence training and extension work to farmers, with its many wide ranging issues, is not discussed. The authors consider that a significant part of the training required for irrigation management staff is the development of an understanding and appreciation of agriculture and the farming community. Simulations and models offer exciting opportunities in training precisely because they are ideally suited to providing such understanding, and are able to link the technical and social features inherent in irrigated agriculture.

THE USE OF SIMULATIONS AND MODELS IN TRAINING

Simulations and models are used in a wide range of situations. Applications may range from scaled down models of large hydraulic structures such as river barrages to sophisticated war games or business management games. Increasingly, simulations and models are being used for technical training, and are often computer based.

The principal features of simulation which make it such a useful training aid can be summarised (after Walford and Taylor 1981) as:-

i) Participants perform under guidance in a non-critical environment. They experience simulated consequences which relate to their decisions and performances.

ii) They monitor the results of their actions and reflect on the relationship between their own decisions and the resultant consequences. If their performance is inadequate they cannot help knowing it and have already done some learning.

iii) Based on their own knowledge and experience participants can make decisions, learn from them, adjust them and progress towards a solution. By developing this process themselves the retention of the knowledge is better assured.

iv) Events and relationships occur and are created in a compressed time period. As a result participants are able to experience, see the consequences of, and learn from situations which would take considerably longer in the real world.

v) Participants are forced to take part and adopt roles. These roles are representations of roles in the real world, and decisions are made in response to the participants' assessment of the setting in which they find themselves.

The use of models in training offer many of the same advantages as simulation. Whereas the part of role playing in the running of a computer-based model is minimal, the model may be used as a more powerful means of conveying a technical or physical message. Running of a model by a participant may take the place of practical hands-on training which is so important a part of learning. As a Chinese aphorism puts it:-

If I hear it, I forget
If I see it, I remember
If I do it, I know.

Simulations and models allow participants an opportunity to "do it" in a speeded-up, non-critical environment where they can assess the results of their actions and their effect on their environment, without fear of the consequences. The two techniques complement one another well in covering the full scope of training requirements for irrigation management. Last, but by no means least, they increase the participants' interest and involvement in the training programme.

EXAMPLES OF SUITABLE SIMULATIONS FOR TRAINING IRRIGATION MANAGEMENT STAFF

One of the authors in cooperation with Dr.Ian Carruthers at Wye College, University of London has developed a training programme for training irrigation management staff incorporating a series of simulations. (Burton and Carruthers 1982). These simulations include:-

(a)   In-Tray Exercise, in which the participant is given an in-tray with various letters, memoranda, and radio-telephone messages which simulate in a concentrated time period the variety of problems that an irrigation system manager might encounter during his working life.

(b)   Main Canal Case Study, in which the participant experiences the possible problems that can occur in the field with control and measurement of irrigation supplies.

(c)   Rotation of Water Supplies, in which the participant has to devise a time-share table for a given head discharge, making due allowances for cropped areas, filling times and losses.

(d)   Scheduling of Crop Planting Dates, in which the participant sub-divides a command area into three or four divisions so that a portion of the area can be planted once the river water supply begins to rise at the start of the season. The remaining areas are subsequently irrigated when there is sufficient supply.

(e)   Prediction of Dry season Flows, in which the participant uses a recession curve to predict the likely dry season water supply and makes recommendations to the water users on the type and area of each crop to grow.

(f)   Irrigation Management Game, in which participants adopt the roles of Operation and Maintenance Assistants or village Water Distribution Officials. The game demonstrates to the participants how directly the actions of these officials affect the livelihood of water users in an irrigated agriculture system. The Irrigation Management Game has been adapted and used on several occasions with undergraduate and postgraduate students at the National College of Agricultural Engineering, Silsoe; the Institute for Irrigation Studies, University of Southampton; the workshop on Irrigation Management Games, Bangalore, India (10-12 January 1983); the Department of Administrative Studies, University of Manchester; and on a Water Resources Development and Management training programme, University of Lund, Sweden.

The training package which includes these simulations has been designed for irrigation systems in East Java, Indonesia with a clearly defined level of participant in mind. The level chosen is that of head of a sub-section who is responsible for the organisation and water distribution for a total command area of approximately 5,000 ha.

A fair degree of knowledge and education on the part of the participants is therefore assumed: obviously the scope and complexity of simulations must be adjusted to take account of the prior knowledge of the participants.

Though the simulations described here were developed for the situation on East Java, many of the principles incorporated within the simulations are relevant to Africa. The package could quite easily be restructured to suit the situation found in tropical Africa.

THE USE OF MODELS IN TRAINING

Models offer similar exciting opportunities in training. The present discussion is confined to the use of digital models, though the use of analog or physical models should not be discounted. Whatever types of model are developed they are potentially extremely powerful training tools; and for the same reason as simulations, that they allow participants to take decisions and to operate the system in a non-critical environment and to experience the effect of different inputs on the performance of the system in a compressed time period. Good models can allow different levels of operation and therefore provide different levels of training. Over a long period they can provide feedback to the scheme operators so that they can relate the model results to those of the real scheme. A particular attraction of the approach of using numerical models in training is the increasing availability, adaptability and ruggedness of micro-computers. Increasing expertise in computer graphics enhances their use as training tools particularly for lower grades of technical staff.

The basic objective of a model for training irrigation management staff is to simulate the operation of a scheme in such a way that the relationship of the various key parameters to one another is accurately represented.

The model can then be run under various ranges of input conditions and the participants (scheme managers and operators) can observe the effect of these variations, and thus gain experience of the decisions which must be taken for successful operation of the scheme. Such a use of a main system simulation has been discussed by Chapman (1981).

There are, of course, already many models which simulate the operation of water resource and irrigation systems. Two noteworthy ones are the Victoria Rivers Model developed by the State Rivers Authority in Australia, which is used as a highly sophisticated aid to operaton in a complex river system, and the Anderson-Maas model developed in America. This latter model simulates decision-making at several levels of activity in an irrigation system, from the operators of the water distribution system to the farmers, and takes into account the response of different crops to irrigation water. Different decision-making activities are assigned to sub-routines in the general program, the essence of which is to parallel the actual distribution and use of irrigaiton water within a system throughout an irrigation season. (Anderson and Maas 1978)

Such models as these, however, run the risk of being too complex for use in training programmes. Firstly the participants may be overwhelmed by the scope of the computer program itself and therefore unable to comprehend its function or its relevance to the everyday reality of the scheme. Secondly a really comprehensive model may include so many variables and interactions that it is impossible to demonstrate fundamental concepts clearly. Thus key features which the training programme requires to demonstrate (for instance the relationship of crop response to water availability and distribution) may rapidly become obscured, particularly for lower levels of management. The great advantage of using models is to show the interactions of the key parameters and this advantage may be lost if too much is attempted within one model.

Figure 14 shows the flow chart for a suitable scheme model for a small irrigation system with pumped supplies from a river. The two way process being represented consists, in one direction, of the calculation of field water requirements for each field unit, allowance for field losses, summation of the field outlet requirements, allowance for canal losses and hence calculation of scheme head requirements. In the other direction the available water must be allocated down the system, making due allowance for losses, and distributed between crops to give the soil moisture balance and crop response required for that period.

Such a Scheme Model can be used both as a training tool and an operating tool so that scheme managers, canal superintendents, agricultural officers and others involved in the operation of the scheme can test it under varying conditions (cropping patterns, irrigation schedules crop priority policy, etc.) and can thus become familiar with the physical response of the system to these variations.

For a comprehensive training programme a set of models is required comprising a Scheme Model, Canal System Model and Agricultural System Model whose relationship to one another is shown by Figure 15. Although the Canal System and Agricultural System Models are in effect sub-routines of the main Scheme Model, they are different from sub-routines in that they each include parameters which are not represented in the Scheme Model. These are parameters which are of a lesser order of importance to the operation of the particular system. Such a parameter in the Canal System Model may be transient flow conditions and sedimentation in canals, while the Agricultural System

135

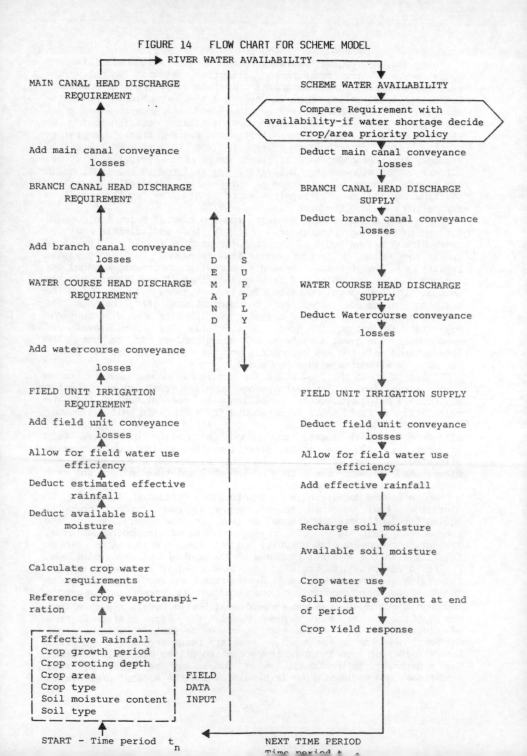

FIGURE 14   FLOW CHART FOR SCHEME MODEL

FIGURE 15

DIVISION OF MODELS FOR IRRIGATION MANAGEMENT TRAINING

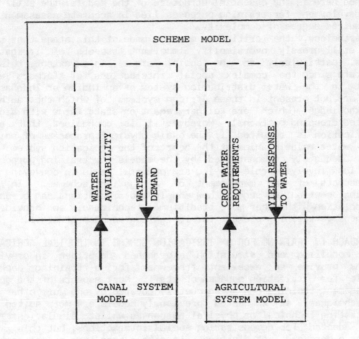

Model might include fertiliser uptake with irrigation supplies and other factors of relevance to the agricultural management of the scheme but only of indirect importance to its overall performance. Of course considerable further complexities exist and could be built in, either to the System Model or to the Scheme Model. What is being discussed here is the essential structure of the models: the skill in their development for training purposes lies in accurate assessment of the required degree of complexity.

Nevertheless, the criticism may be made at this stage that such models still grossly oversimplify important aspects of irrigation systems, particularly on the social side. To an extent, this is deliberate since the complex social interactions of farmers with respect to the water distribution system of an Indian or Indonesian system are not present in those African systems of which the authors have knowledge (which are either tenant or state farms with highly regulated cropping patterns). Moreover it is emphasised that such simplification is required if the basic physical processes of soil - plant - water relationships at the heart of the irrigation system are to be adequately demonstrated by the models for training purposes. Simulations involving role-playing are powerful tools in conveying the human aspects of the management of irrigation schemes. In the following section we suggest how models and simulations can be used, in association with other more traditional techniques, to provide a comprehensive training package for irrigation management staff.

AN APPROACH TO TRAINING FOR AN IRRIGATION SCHEME IN TROPICAL AFRICA

The modelling and simulation approaches described in previous sections provide an excellent framework for a training package suitable for irrigation management staff. Most aspects of the scope of management training can be adequately covered by one or other of the techniques; as suggested previously models are more suited for demonstrating technical or physical responses whilst simulations are better adapted for demonstrating social interactions, but this is by no means a hard and fast division. In either case the normal method of use would be to run the simulation or model early in the training programme so that the participants can put each element of the training in overall perspective. The course then consists of lectures, seminars, practicals, audio-visual aids and other methods as appropriate. Finally the models and simulations are run again at the end of the course to reinforce the training. Throughout, the emphasis is on learning rather than teaching, an approach assisted by the use of models. Each model has certain applications for certain topics at certain levels of expertise and the approach must be modified for each training situation. The important features of each approach and its relevance to the overall training programme are discussed briefly below:

(a) The Scheme Model
This is the basic package for all participants. It shows how the major inputs interact but it is of particular value to senior scheme management in allowing them an overview of scheme operation. Major features covered under this model are:

- scheme design and operation
- cropping patterns and water use
- principles of water allocation and distribution
- yield of response to water
- monitoring and evaluation

Treatment of these matters in the subsequent training package would be mainly conceptual in approach for this relatively senior group. Detailed exercises where necessary would be covered by the other models.

(b) The Canal System Model
This model forms the framework for the package for all those involved in the operation of the canal system. The main external inputs are the water requirements at the field outlets and the water availability at the system head. The model simulates the actual operation of the canal system. Major features covered under the model include:

- system layout
- channel levels and discharges
- backwater effects
- regulator operation
- conveyance efficiencies
- principles of water allocation
- rainfall
- drainage
- flood control
- sedimentation
- maintenance

Lectures, reinforced by exercises, form the basis of the subsequent training programme. Much of this training programme is very suitable for question-and-answer routines on visual display units where computer graphics can be a particularly powerful aid in training on such aspects as system layout.
The Canal system package can be used in appropriate parts by all levels of management from the scheme managers through the canal supervisors to the gate operators.

(c) The Agricultural System Model
This forms the framework for the agricultural training package. Further development is required by specialists, but it appears that simplicity may require that the model be divided into two parts:-

i)   the input model
ii) the output model

The input model would cover crop water requirements, labour requirements, agricultural inputs and basic agronomy and the model could be used to test various cropping patterns in respect of these parameters and to demonstrate to participants their interaction. The output model would cover basically principles of water allocation, field irrigation methods and crop response to water. Soil - plant - water relationships would be a basic feature subsumed in both models.

As in the case of the Canal System package, subsequent training would take the form of lectures and exercises. Exercises on, say, crop response to water would be essential to ensure that the matter was properly understood and had not been obscured by computing techniques. Many of the aspects covered by the Agricultural System Model can of course be most convincingly demonstrated by practical means.

(d) Role-Playing Simulation

Role-playing simulations, similar to the Irrigation Management Game, form a vital part of the overall training approach. They provide an integrative function in bringing together the administrators, engineers, agriculturalists and others who are responsible for operating the scheme. By simulating processes and adopting roles the participants are able to obtain a clearer understanding of the complex interactions of the irrigation scheme, particularly with regard to such aspects as the need for communication. They are able to discuss the results and agree communally on the major constraints and requirement for action. Whilst it would be time consuming to develop a game suitable for every individual scheme, the development of a game suitable for application over wide areas of tropical Africa would be relatively easy.

In a slightly different, though related, vein the authors are developing a computer-based design and operation simulation for use by designers of irrigation systems. The objective of the simulation is for designers to design a system and then operate it for a simulated five year period within the constraints of their design parameters. This exercise encourages them to think more actively of the operational procedures involved in irrigation systems and to design accordingly. The data base of the simulation is a 2,000 ha project in East Africa which has passed through feasibility and design stage and is currently under construction. The scheme has surface and sprinkler irrigation for cotton, maize, sugar cane and soya bean crops, and forms a sufficiently small and straight forward layout with which to perform the exercise.

FUTURE DEVELOPMENT

The use of models and simulations, though already well established in many other disciplines, is still in its early stages for the training of irrigation management staff, and future developments are unclear. Without doubt increasingly sophisticated irrigation models will be constructed. For example, the authors are associated with a computer-based Irrigation Management Game being developed under the auspices of Wye College, University of London. This will be similar in concept to the highly successful Wye Farm Game (the Wye Farm Game formed the basis of the annual United Kingdom farmscan competition sponsored by ADAS and Barclays Bank Ltd. between 1979 and 1981, Youngman 1974) and will involve high level administrators, planners and academics from many different countries "playing" the game simultaneously, sending and receiving results by post. Such a game is likely to be conceptual in nature, rather than representative of a particular scheme. It is also likely to be based on an East Asian rice system model and hence not be directly applicable to African systems. Nevertheless participation in the game could be of great indirect benefit to this level of management in Africa since many of the concepts (for instance water distribution and losses, timeliness of supply and crop response to water) are of universal applicability.

SUMMARY AND CONCLUSIONS

The relevance and application of simulations and models incorporated in training programmes for irrigation management staff have been outlined. These techniques for encouraging participants to learn for themselves and to appreciate different points of view are as yet little developed in this field. The authors have experience of the development and use of simulations and models and feel that they can be a valuable tool with which to tackle the multi-disciplinary problems encountered in irrigate agriculture. Their future development and application is expected.

REFERENCES

Anderson, R.L. and Maas, A. (1978). A simulation of irrigation systems. Technical Bulletin No. 1431, Department of Agriculture, Washington D.C.

Burton, M.A. and Carruthers, I.D. (1982) Irrigation management - A training programme incorporating simulations, games and role-playing exercises. Wye College/Sir M. MacDonald & Partners Ltd., Cambridge.

Chapman, G.P. (1981) Gaming simulations of irrigation systems - prospects for management training. Ford foundation, New Delhi, India.

Walford, R. and Taylor, D. (1981) Learning and the simulation game, Open University Press.

Youngman, J.P. (1974) The Construction, Implementation and appraisal of a Farm Management Game. Unpublished, M. Phil. Thesis, Wye College, University of London.

EPILOGUE: PRESPECTIVES ON AFRICAN IRRIGATION

W.M. Adams and A.T. Grove

INTRODUCTION

These papers represent the formal part of the proceedings at the workshop in Cambridge in March 1983. An equally important element was the discussion and informal contributions to the debate. No attempt was made at the workshop to establish a consensus, or to draw conclusions to which all participants could subscribe. Even so, it is important to draw together some of the recurrent threads of debate as a contribution to future work. Most of the ideas and conclusions discussed here were put forward by participants in the workshop. They do not, however, constitute a summary or consensus of views. They are essentially our personal views.

STUDYING POOR PERFORMANCE

Most irrigation schemes in Africa, particularly those in Nigeria, have performed poorly. There is a worrying gap between predicted and actual performance. Schemes which were convincing on paper look dismal in practice, beset by problems of low participation by farmers or even open opposition from them, slow expansion of irrigated cropping, and disastrously low rates of return. It is easy to categorise the majority of schemes as 'failures' to some extent, and to castigate developers with their inadequacy. As Henry (1978) put it:

> "we don't fly in airplanes (sic) with a 50% failure rate.
> Villagers don't want machines which can't be repaired".
> 1978:365

By the same token, failed irrigation schemes are worse than useless.

But is this conclusion of almost universal failure valid? Quite apart from the potentially unhelpful and pejorative overtones of the word "failure", there are a number of reasons for caution. The first is that it is too easy to draw sweeping conclusions from a few case studies, and there are still too few adequate accounts of irrigation schemes in Africa. There may be marked regional variations: the lessons of Nigeria, for example, may not necessarily be applicable to the rest of the continent. Second, any discussion of scheme failure must start by setting the criteria of judgement, and these can vary. Some of the schemes described at the workshop seem to have performed badly in anyone's language, but a more detailed definition of the objectives of irrigation development, and value-judgements about its aims and the interests it ought to be serving, are needed if relevant criteria to judge success or failure are to be established. Schemes can meet, or ostensibly attempt to meet, objectives of social welfare, rational economics or commercial interests, or can serve political ends at various levels. Unless there is a clear understanding of who is trying to get what out of an irrigation scheme, then words like "success" or "failure" are as flexible and weak as the nebulous "development" the schemes are probably being marshalled to achieve.

More field studies of irrigation schemes and more reliable data are required urgently if the debate about irrigaion is to progress. These studies must be firmly based in the practical experience of firms, development agencies and governments, and above all their results must be disseminated and discussed. There must be a lot more integration of research and policy discussions, or else academics will continue to criticise from the sidelines while practitioners plough on regardless, seemingly deaf and blind to the problems. Some argue that social scientists are too ready to pounce on new initiatives in development and highlight their initial failures. After all, as Chambers (1980) comments: "we earn more points professionally by writing about failure than about success" (p. 96). It is probably true that academics are too quick and too sweeping in their criticisms of irrigation, and not well-enough informed about practicalities. The focus on "failure" is, however, potentially very productive, since the problems highlighted are serious.

FAULTY PLANNING

A number of the problems on African irrigation schemes can be attributed to shortcomings of the project planning process. The papers on soil survey alone show how many pitfalls there are even in this time-honoured and highly technical input to planning. The practical difficulties Wayne Borden outlines in this volume will be familiar to anyone who has had experience in the field, but the theoretical limitations of the survey methods - their sheer inappropriateness to either developer or developed - described by David Dent and John Aitken truly give one pause.

Frequently, and to a serious extent, the cart is being put before the horse in irrigation planning. Indeed, the problem is common to all development planning. Political structures and means of communicaion that would enable the needs and desires of local people to be expressed at the national or even the project management level are usually inadequate. Development projects are consequently being imposed from above, with all the attendant agonies of trying to fit square pegs (traditional farmers and established farming and socio-economic systems) into round holes in the shape of the new irrigated cropping patterns and lifestyles. There are rarely adequate socio-economic surveys at the early stage of irrigation planning. If done at all, they are perfunctory, stereotyped and late in the planning process. Their results are not integrated into planning important fundamental constraints on what is or is not feasible, but they are generally used instead as a kind of ornament, or at best fine tuning of plans drawn up on economic or engineering criteria. Too often socio-economic surveys are used to reveal the source of likely opposition to irrigation so that appropriate extension techniques can be developed, rather to examine the roots of and reasons for those attitudes.

Obviously, the idea of development from below (cf. Pitt 1976) is attractive in theory, but can it be applied in practice? There is evidence that it can; even the more technical elements of surveys for project planning such as soil survey can be usefully influenced by the consideration at the earliest stages of planning of needs of local people, the projects' future participants, as the work of the War on Want Project in Mauretania shows (Bradley et al. 1977, Bradley 1980, 1983).

On a more general scale, all irrigation projects should start with a survey of the perceived requirements of the local community, rather than simply canvassing opinion on one scheme or on a small group of pre-planned options for development. This demands that social science becomes the lead discipline in development planning. This is a major reversal of roles vis-a-vis the established disciplines of engineering, agronomy, soils and economics and is likely to be a painful transformation. Nonetheless, if it is the social and economic aspects of schemes which are pre-eminently responsible for poor performance, they cannot simply be planned away by more socio-economic studies, there has also to be a shift in emphasis in the established procedure of project planning. The corollary of this, of course, is that the social sciences must seek procedures which can convert raw information, however interesting, into a practical land management system that will work. In other words, the established irrigation planning disciplines have to give the social sciences a chance; the latter must then come up with solutions.

EVALUATION

More evaluation studies of irrigation schemes in tropical Africa are needed. Too little is known about the performance of schemes, certainly outside the agencies and companies immediately involved, and there is little dissemination of experience which can be applied to new schemes. Three sorts of evaluation studies seem to be needed. First, internal monitoring of scheme performance is important. The "information problem" is a significant factor in poor performance (Wade 1981) and there is clearly a future for action research combining evaluation with trying out solutions to revealed problems (cf. Bottrall 1981). Second, independent evaluation is needed, and comparative studies of different schemes, to put scheme performance into a wider context. Third, there is a need for retrospective studies of established working and defunct schemes, using archival or oral sources to establish how they worked and where problems occured.

Of these three the last is perhaps the least considered and the most immediately promising. It is a field where academic research may be able to make a valuable contribution, and the considerations of responsibility for shortcomings may be less acute. It is also possible that retrospective studies of management and performance could be integrated into the technical rehabilitation work which is seemingly a growing feature of irrigaition scheme design and management. In the long run, however, it is the integration of evaluation at successive stages which will be the most important. It is vital that mechanisms be developed for proper assessment, and for aborting schemes which go astray.

APPROPRIATE DEVELOPMENT

Consideration of irrigation schemes in tropical Africa raises a series of questions about the speed and the scale of development. The rate at which new schemes have been conceived, and the speed with which they have been planned and executed in the last decade, must be seen as factors contributing to the lack of success some of them have suffered. The construction of the Bakolori Project in Nigeria for example was deliberately rapid, an attempt to offer "instant development" ostensibly cutting out the delay and inefficiency of slower developments. That very speed of development left little or no room for manoeuvre once shortcomings in planning and design over questions of land in particular were revealed (Bird, this volume and Adams, this volume). Rapid development of individual schemes accentuates problems of implementation and management. Rapid development of irrigation on a wider scale can lead to shortages of skilled manpower, and certainly does not help the process of careful evaluation and dissemination of experience that would seem to be necessary to make irrigation a success.

There are questions also about appropriate size in irrigation schemes. Small-scale schemes have certainly become fashionable recently, and are being promoted by the World Bank, among others. To an extent this may be a fashion, the supposed benefits unproved by case studies. Nonetheless, they would seem to have several real advantages over large-scale schemes:

"the strongest argument in favour of small-scale irrigation is that it is easier than large-scale development because the human problems are reduced to a manageable scale"

(Stern 1979, p. 23)

In addition there would seem to be more chance of trying out "bottom-up" approaches to development, that is the involvement of farmer participation, in small-scale schemes. Problems of corruption, endemic on large-scale schemes, may be reduced on smaller ones.

The environmental impact of schemes must not be neglected. Disease hazards, the availability of fuelwood (Hughes, this volume), and salinity and drainage problems can add enormously to costs in the long term and yet they receive inadequate attention in the planning process, in the course of project implementation and monitoring, and from researchers in Africa.

THE ROLE OF CONSULTANTS

The role of consultants, construction companies and individual operators is crucial to the study of irrigation schemes. In Africa the majority of pre-feasibility, feasibility, design, construction, supervision and evaluation studies are made by expatriate consultants. It is not unusual for consultants to have access to, and to be equipped to make use of, more data, more technical expertise, and even a wider planning perspective than the client government or agency. In such circumstances, the consultant's approach to the development problem and the solution, or range of solutions, he offers will to a large extent predetermine the decision taken on the development. In other cases, terms of reference may be prepared which constrain the development options so that the consultant's decision to accept them means that he finds his own work too narrowly constrained, and desirable options that become apparent are outside his mandate.

Either way, the consultant is a key factor in determining the direction of development. Many of the less successful schemes in Africa have enjoyed or suffered from the involvement of consultants at successive stages: it is sobering that the problems with the schemes were either not recognised or, more disturbingly, not pointed out at an earlier stage. The reason for this is not in general any lack of expertise, though myopia and amnesia do seem to beset irrigation planning; the problem is that the consultant is not free, as the academic is too prone to assume, to do as he wishes. The constraints of commercialism and the demands of contracts should not be underestimated. The effects exaggerate the myopia and amnesia, and can contribute to poor irrigation schemes. Irrigation planning is in part an academic exercise, but for a consultant it is business. "We bring solutions or promises of solutions and exchange them for money", writes Watson (1981, p. 8). In this respect it should be noted that the commercial company is no more constrained than the academic out in the world selling his skills (Penning-Rowsell, 1981).

There comes a point where professional concern overrides commercial considerations, but that point is not always clearly recognisable to more than a few individuals. One approach to a new order in irrigation consultancy might require consultants not to engage in a project preparation exercise unless socio-economic surveys have ascertained local community concerns, and unless there is provision for proper maintenance and staff training. The notion of socially responsible research (Preston, 1982) might be applied to consultancy with good effect.

BEING WRONG AND BEING REALISTIC

Commercial factors and the constraints of confidentiality greatly limit our freedom and willingness to accept our mistakes. In the field of irrigation:

"we worship intelligence, but far too often intelligence is thought of as the quality of avoiding being demonstrably wrong about anything" (Chambers 1977, p. 111).

Very often, as Robert Chambers points out, the most significant insights into what is really going wrong with an irrigation scheme are kept confidential: the studies are made but they are marked "not for quotation" (Chambers, 1981 p. 8). Obviously, honesty about poor scheme performance, and a forum where such things can be discussed, are prerequisites for improved performance.

Other elements, some of them equally painful, are also important in the learning process. One is the humility to admit how little we really know. Fyfe (1980) has described the repeated efforts of successive governors and experts to develop agriculture in Sierra Leone in the nineteenth century:

"plainly it appears that once experts' insights are firmly

entrenched they need no aid from empirical evidence - indeed they can lightheartedly ignore empirical evidence entirely" (Fyfe 1980 p. 11).

There are barriers between disciplines, barriers between academic and consultant, and above all barriers between expatriate 'expert' of whatever camp and African government and agency. All must be dismantled before there is much hope of tackling with any success the task of cooperating with African farmers to increase agricultural productivity by means of irrigation. One of the hard questions raised by the workshop was whether we are in fact, for all the experience that has been gained, getting any better at planning, building, or managing irrigation schemes. It is a disturbing thought that we may not.

REFERENCES

Bradley, P.N. (1980) Agricultural development planning in the Senegal valley, in E.S. Simpson (Ed.) The rural-agricultural sector. Institute of British Geographers Developing Areas Study Group.

Bradley, P.N., C. Reynaut & J. Torrealba (1977) The Guidimaka region of Mauritania: a critical analysis leading to a development project. War on Want Press, London.

Bottrall, A. (1981) Improving canal management: the role of evaluation and action research. Water Supply and Management 5: 67-79.

Chambers, R. (1971) Challenges for rural research and development, p. 398-412 in B.H. Farmer (Ed.) Green revolution? Macmillan.

Chambers, R. (1980) Cognitive problems of experts in rural Africa, p. 96-102 in J.C. Stone (Ed.) Experts in Africa. African Studies Group, University of Aberdeen.

Chambers, R. (1981) In search of a water revolution: questions for canal irrigation management in the 1980s. Water Supply and Management 5,5-18.

Fyfe, C. (1980) Agricultural dreams in nineteenth century West Africa, p. 7-11 in J.C. Stone (Ed.) Experts in Africa. African Studies group, University of Aberdeen.

Henry, D. (1978) Designing for development: what is appropriate technology for rural water and sanitation? Water Supply and Management 2: 365-72

Penning-Rowsell, E.C. (1981) Consultancy and contract research, Area 13: 10-12

Pitt, D. (1976) The Social Dynamics of Development. Pergamon Press, Oxford.

Preston, D.A. (1982) Communities, individuals and research. Applied Geography 2: 307-312.

Stern, P.H. (1979) Small-scale irrigation: a manual of low-cost water technology. International Technology Publishing Company, London.

Wade, R. (1981) The information problem of South Indian irrigation canals. Water Supply and Management 5: 31-51.

Wade, R. (1982) The system of administrative and political corruption: canal irrigation in South India. J. Development Studies 18: 287-328.

Watson, R.M. (1981) Down market remote sensing. p. 5-63 in the Proc. of the 9th Annual Conference of the Remote Sensing Society, Matching remote sensing technologies and their application. London, December 1981.